High Energy Particle Physics Workshop (HEPPW 2015)

Journal of Physics: Conference Series Volume 645

Johannesburg, South Africa
11 - 13 February 2015

ISBN: 978-1-5108-1472-1
ISSN: 1742-6588

Printed from e-media with permission by:

Curran Associates, Inc.
57 Morehouse Lane
Red Hook, NY 12571

Some format issues inherent in the e-media version may also appear in this print version.

Printed by Curran Associates, Inc. (2016)

For permission requests, please contact the Institute of Physics
at the address below.

Institute of Physics
Dirac House, Temple Back
Bristol BS1 6BE UK

Phone: 44 1 17 929 7481
Fax: 44 1 17 920 0979

techtracking@iop.org

Additional copies of this publication are available from:

Curran Associates, Inc.
57 Morehouse Lane
Red Hook, NY 12571 USA
Phone: 845-758-0400
Fax: 845-758-2634
Email: curran@proceedings.com
Web: www.proceedings.com

Table of contents

Volume 645

High Energy Particle Physics Workshop (HEPPW2015)
11–13 February 2015, Johannesburg, South Africa

Papers

Theoretical collider

Experimental collider

Instrumentation

Preface

The motivation for this workshop began with the discovery of the Higgs boson three years ago, and the realisation that many problems remain in particle physics, such as why there is more matter than anti-matter, better determining the still poorly measured parameters of the strong force, explaining possible sources for dark matter, naturalness etc. While the newly discovered Higgs boson seems to be compatible with the Standard Model, current experimental accuracy is far from providing a definitive statement with regards to the nature of this new particle. There is a lot of room for physics beyond the Standard Model to emerge in the exploration of the Higgs boson. Recent measurements in high-energy heavy ion collisions at the LHC have shed light on the complex dynamics that govern high-density quark-gluon interactions. An array of results from the ALICE collaboration have been highlighted in a recent issue of CERN courier. The physics program of high-energy heavy ion collisions promises to further unveil the intricacies of high-density quark-gluon plasma physics.

The great topicality of high energy physics research has also seen a rapid increase in the number of researchers in South Africa pursuing such studies, both experimentally through the ATLAS and ALICE colliders at CERN, and theoretically. Young researchers and graduate students largely populate these research groups, with little experience in presenting their work, and few support structures (to their knowledge) to share experiences with. Whilst many schools and workshops have sought to educate these students on the theories and tools they will need to pursue their research, few have provided them with a platform to present their work. As such, this workshop discussed the various projects being pursued by graduate students and young researchers in South Africa, enabling them to develop networks for future collaboration and discussion.

The workshop took place at the iThemba Laboratories - North facility, in Gauteng, from the 11th to the 13th of February 2015, where excellent conference facilities with outdoors and indoor tea areas for discussions and interactions were provided, along with a state-of-the-art remote access to the conference venue such that those who were unable to attend the workshop in person could also be present. The laboratory is located next door to the Wits Professional Development Hub (on the corner of Jan Smuts Avenue and Empire Road), which provided the catering for this workshop. A morning plenary session, followed 15+10 minute presentations, was the format across our three days. The topics covered being in high-energy theory and phenomenology (heavy ions, pp, ep, ee collisions), ATLAS physics and ALICE physics. The workshop website is http://hep.wits.ac.za/HEPPW2015.php

Alan S. Cornell and Bruce Mellado

Journal of Physics: Conference Series **645** (2015) 011001 doi:10.1088/1742-6596/645/1/011001

Figure 1. *The opening address delivered by Zeblon Vilakazi.*

Figure 2. *The welcome delivered by John Carter.*

Journal of Physics: Conference Series **645** (2015) 011002 doi:10.1088/1742-6596/645/1/011002

Sponsors

- National Institute for Theoretical Physics (NITheP)
- The University of the Witwatersrand
- iThemba Laboratories
- Mandelstam Institute for Theoretical Physics

Organising Committee

Bruce Mellado	University of the Witwatersrand - (*co-chair*)
Alan S Cornell	NITheP; University of the Witwatersrand - (*co-chair*)
Elias Sideras-Haddad	Materials Physics Research Institute; University of the Witwatersrand
Zinhle Buthelezi	iThemba Laboratories
Will Horowitz	University of Cape Town
Andrew Hamilton	University of Cape Town
Deepak Kar	University of the Witwatersrand
Sahal Yacoob	University of Cape Town

Journal of Physics: Conference Series **645** (2015) 011002 doi:10.1088/1742-6596/645/1/011002

List of speakers

Ammar ABDALGABAR	University of the Witwatersrand
Ryan ATKIN	University of Kwa-Zulu Natal
Warren CARLSON	University of the Witwatersrand
John CARTER	University of the Witwatersrand
Alan S CORNELL	NITheP; University of the Witwatersrand
Mitchell COX	University of the Witwatersrand
Tom DIETEL	University of Cape Town
Jonathan ESTEVES	University of the Witwatersrand
David GOSSMAN	University of the Witwatersrand
Guillermo HAMITY	University of the Witwatersrand
Gerhard HARMSEN	University of the Witwatersrand
Harshna JIVAN	University of the Witwatersrand
Deepak KAR	University of Glasgow
Thokozani KHUMALO	University of the Witwatersrand
Isobel KOLBE	University of Cape Town
Mukesh KUMAR	University of the Witwatersrand
Oscar KUREBA	University of the Witwatersrand
Claire LEE	University of Johannesburg
Shell-May LIAO	University of the Witwatersrand
Lucas McCONNELL	University of Cape Town
Luis MARCH RUIZ	University of the Witwatersrand
Ben MEIRING	University of Cape Town
Bruce MELLADO	University of the Witwatersrand
Razieh MORAD	University of Cape Town
Simon MULLINS	iThemba North
Azwinndini MURONGA	University of Johannesburg
Chad PELWAN	University of the Witwatersrand
Daniel OHENE-KWOFIE	University of the Witwatersrand
Robert REED	University of the Witwatersrand
Xifeng RUAN	University of the Witwatersrand
Marc SACKS	University of the Witwatersrand
Ferdinand SCHENCK	University of Cape Town
Joshua SMITH	University of Cape Town
Matthew SPOOR	University of the Witwatersrand
Xolisile THUSINI	University of Cape Town
Kehinde TOMIWA	University of the Witwatersrand
Zeblon VILAKAZI	University of the Witwatersrand
Stefan VON BUDDENBROCK	University of the Witwatersrand

Peer review statement

All papers published in this volume of *Journal of Physics: Conference Series* have been peer reviewed through processes administered by the proceedings Editors. Reviews were conducted by expert referees to the professional and scientific standards expected of a proceedings journal published by IOP Publishing.

Large trilinear A_t soft supersymmetry breaking coupling from 5D MSSM

Ammar Abdalgabar[1,†,‡] **and A. S. Cornell**[2,†]

† National Institute for Theoretical Physics; School of Physics and Mandelstam Institute for Theoretical Physics, University of the Witwatersrand, Johannesburg. Wits 2050, South Africa
‡ Department of Physics, Sudan University of Science and Technology, Khartoum 407, Sudan

E-mail: [1]`ammar.abdalgabar@students.wits.ac.za`, [2]`alan.cornell@wits.ac.za`

Abstract. The possibility of generating a large trilinear A_t soft supersymmetry breaking coupling at low energies through renormalisation group evolution in the 5D MSSM is investigated. Using the power law running in five dimensions and a compactification scale in the 10-10^3 TeV range, we show that gluino mass may drive a large enough A_t to reproduce the measured Higgs mass and have a light stop superpartner below ~ 1 TeV, as preferred by the fine tuning argument for the Higgs mass.

1. Introduction

The observed scalar particle in 2012 of mass ~ 125.5 GeV [1, 2] is consistent with the Standard Model (SM) Higgs boson. In the context of the Minimal Supersymmetric Standard Model (MSSM), this motivates considering models of supersymmetry breaking in which the stop superpartner is heavy (beyond the reach of the LHC) or a model in which a large trilinear A_t soft supersymmetry breaking parameter can be generated at low energies. The first option is disfavoured by fine-tuning arguments, while the second one allows for lighter stops ~ 1 TeV, and it is thus preferred by naturalness arguments. The solution advocated in this work is to resolve the previous issue by introducing an extra (fifth) dimension and taking advantage of the power low running to generate a sizable value of A_t starting from a very small value.

We define 5-dimensional (5D) MSSM as where the Higgs superfields and gauge superfields always live in the bulk. As a consequence these fields will have Kaluza-Klein modes which contribute to the Renormalisation Group Equations (RGEs) at $Q > 1/R$ and additional matter associated to five dimensional $\mathcal{N} = 1$ super Yang-Mills. Different possibilities of localising the matter superfields can be studied. We shall consider the limiting case of superfields with SM matter fields restricted to the brane, and the RGEs for this scenario can be found in Ref. [14]. Therefore there will be no additional Kaluza-Klein contributions of these matter fields to the RGEs[14, 6].

Regarding the breaking of supersymmetry, whilst gauge mediation is favoured (and some recent work on Gauge Mediated Supersymmetry Breaking (GMSB) in a five dimensional context may be found in Ref. [8]), ultimately the universality of squark massses in GMSB means that even though the gaugino mediated limit [14] might allow for light squarks (and 5D RGE evolution allows for a large A_t and the observed Higgs mass), the collider bounds on first and second generation squarks [14], in the supra-TeV range would apply also to the 3rd generation

Journal of Physics: Conference Series **645** (2015) 012001 doi:10.1088/1742-6596/645/1/012001

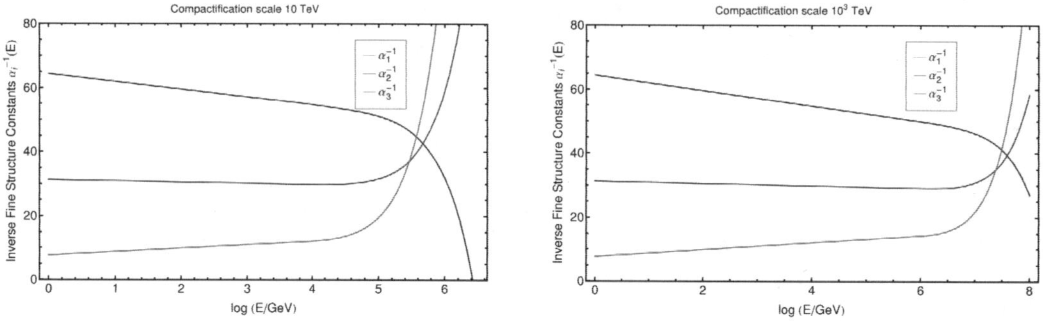

Figure 1. Evolution of the inverse fine structure constants $\alpha^{-1}(E)$, for two different values of the compactification scales 10 TeV (left panel), 10^3 TeV (right), with M_3 of 1.7 TeV, as a function of log(E/GeV).

squarks, i.e. the stops, which as discussed before, is both phenomenologically less interesting and unnatural. Therefore we wish for some other description of supersymmetry breaking that may allow for stops to be lighter than their first and second generation counterparts, such as in Refs. [4, 5]. In this work we will therefore be unspecific about the precise details of how supersymmetry is broken, and as a result our conclusions will also apply quite generally. We do however make some minimal specifications:

- We take as inputs the Yukawa and gauge couplings at the SUSY scale, 1 TeV.
- We will assume supersymmetry breaking occurs at the unification scale, which is found by finding the scale at which $g_1 = g_2$, which is lowered compared to the 4D MSSM, by the effects of the compactification.
- We specify the value of the gluino mass, M_3 at 1 TeV.
- We take the trilinear soft breaking terms, $A_{u/d/e}$, vanish at the unification scale.

We solved the combined set of differential equations numerically by using the above conditions, taking the "third family" approximation in which we only evolve third generation RGEs. This approximation is quite standard and is due to the relative smallness of the other Yukawa couplings (at least one order of magnitude) compared to those of the third generation, and as a result the other A-term values are also very small. We further specified some parameters such as μ, B_μ and the value of the sfermion masses ($\sim$ 1 TeV) so as to allow for the RGEs to be solved, but these do not affect the overall result. We solved the differential equations between $Q_{min} = 10^3$ GeV and Q_{max}, which was typically only one order larger than the unification scale.

A sufficiently large trilinear A_t soft supersymmetry breaking parameter allows us to reproduce the measured Higgs field mass of $\sim$ 125.5 GeV, while keeping a light stop superpartner (below $\sim$ 1 TeV) as preferred by the fine tuning argument for the Higgs mass. Realising such a high A_t is usually difficult (in supergravity, mGMSB, etc.) it is shown here concretely that 5D MSSM with a compactification scale around 10-10^3 TeV can achieve a large A_t at low scale (starting from $A_t(M_{GUT}) \sim 0$), thanks to the power law running; and simultaneously with an approximate unification of the gauge couplings, the precision of this gauge unification is illustrated quantitatively (at one-loop) for various values of the compactification scale as pictured in Fig. 1. We also specify the Yukawa coupling RGE [14] boundary conditions at 1 TeV, which interestingly appears to vanish when evolved to the unification scale as shown in Fig. 2.

More precisely, it is found that increasing the compactification radius tends to increase the size of the trilinear A_t parameter at low scale as shown in Fig. 3. In fact the absolute value

Journal of Physics: Conference Series **645** (2015) 012001	doi:10.1088/1742-6596/645/1/012001

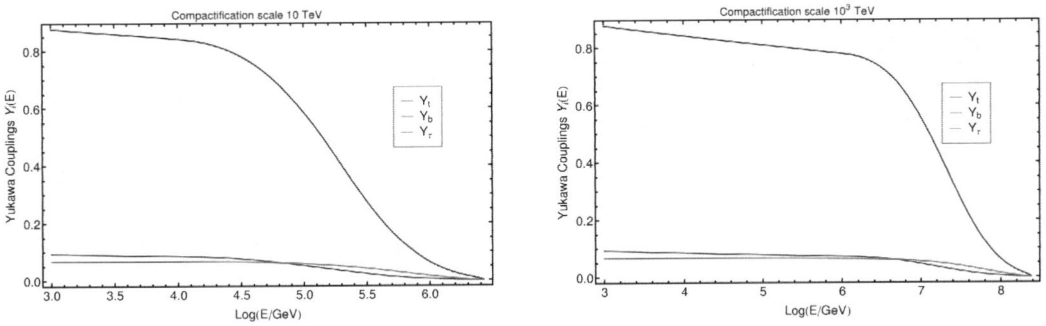

Figure 2. Evolution of Yukawa couplings Y_i, for two different values of the compactification scales: 10 TeV (left panel), 10^3 TeV (right), with $M_3[10^3]$ of 1.7 TeV, as a function of log(E/GeV).

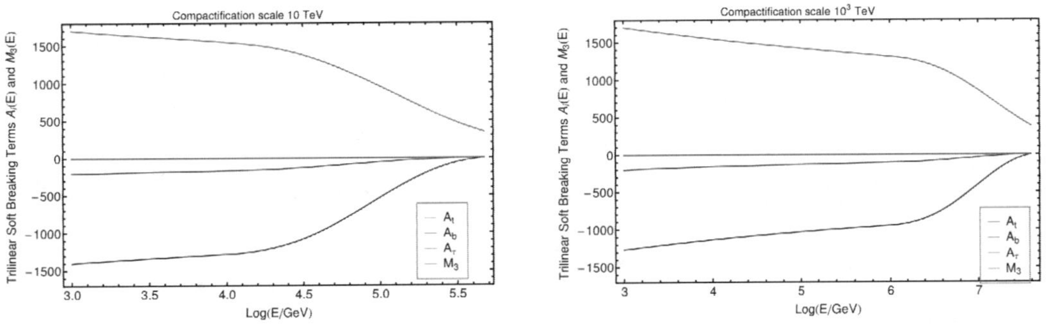

Figure 3. Evolution of trilinear soft terms $A_i(3,3)(E)$, for two different values of the compactification scales 10 TeV (left panel), 10^3 TeV (right), with $M_3[10^3]$ of 1.7 TeV, as a function of log(E/GeV).

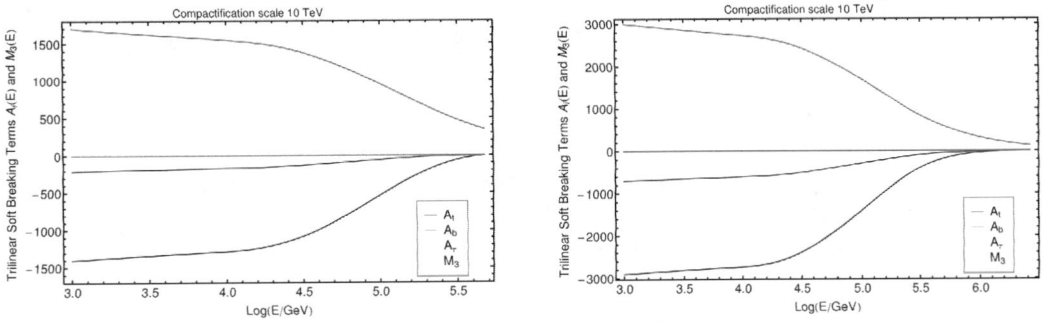

Figure 4. Evolution of trilinear soft terms $A_i(3,3)(E)$, for two different values of gluino masses, M_3: 1.7 TeV (left panel), 3 TeV (right panel), with R^{-1} of 10 TeV, as a function of log(E/GeV).

of A_t mimics the magnitude of the final value of the gluino mass (at $1/R$) as can be seen in Fig. 4. As this interesting result of large A_t generated radiatively is rather generic, the origin of supersymmetry breaking is left unspecified and the treatment adopted is effective.

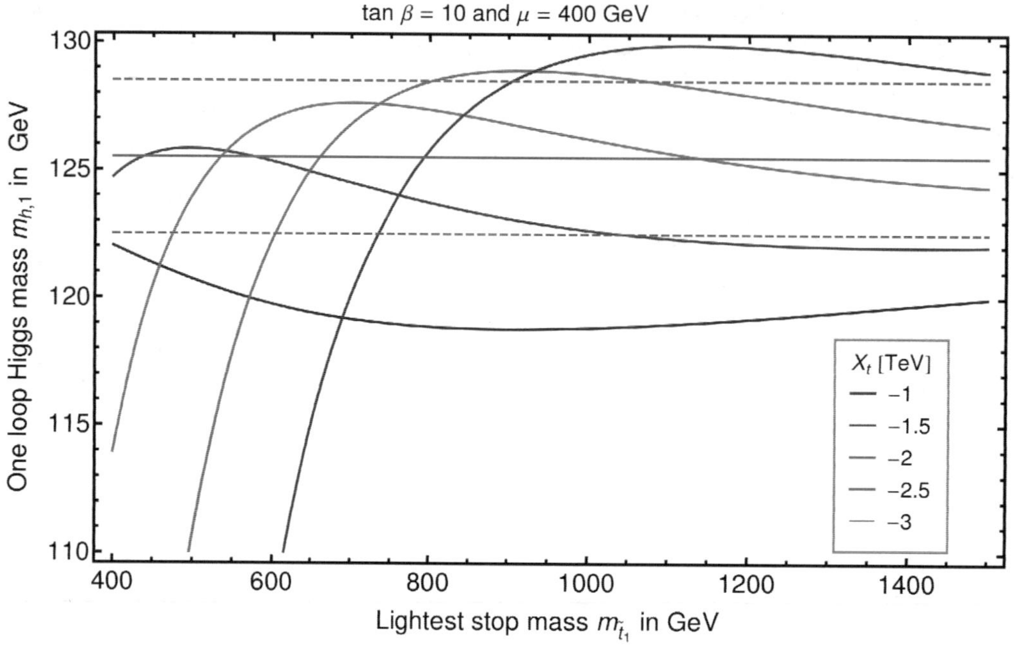

Figure 5. One-loop Higgs mass versus the lightest stop mass for representative values of $X_t = A_t - \mu \cot\beta$, corresponding to those of the 5D MSSM.

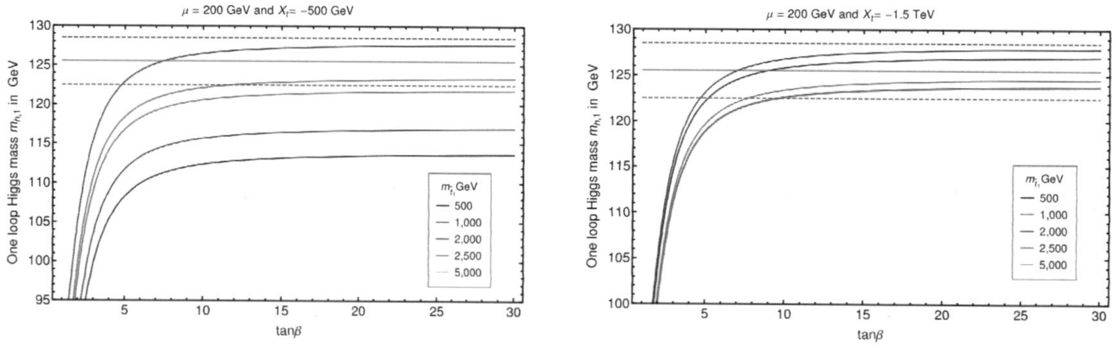

Figure 6. One-loop Higgs mass versus $\tan\beta$ for different values of the stop mass, for $X_t = A_t - \mu \cot\beta$ of -500 GeV (left panel) and -1.5 TeV (right panel).

This large A_t can be realised in both cases of the three quark generations being on the boundary or only the third generation of matter fields on the boundary ("split families"). In the later case, the stop can be much lighter and possibly fall inside the LHC reach. All these results are based on the beta function for the gauge couplings, Yukawa couplings, trilinear terms, scalar soft masses, Higgs masses, μ and B_μ parameters, whose analytical expressions can be found in Ref. [14].

Using the one-loop Higgs mass formula in Ref. [14] and fixing $m_{h,1} = 125.5$ GeV, $m_Z = 91$ GeV, $\mu = 200$ for $\tan\beta = 10$ we can see in Fig. 5 that for representative values of A_t, achievable in the 5D MSSM, one may easily accommodate the lightest stop mass in the sub-TeV range.

Let us pay attention to dependence on the value of $\tan\beta$ as pictured in Fig. 6. In fact the value of $\tan\beta$ will depend greatly on how μ and B_μ are addressed in the context of supersymmetry breaking and hence the solution of the vacuum tadpole equations, but regardless of this, for values of $\tan\beta > 10$ the functions are approximately flat and we expect the value to fall within this interval. We expect that the μ term is naturally of the order of the electroweak scale, where in Fig. 5 we took a slightly large μ value of 400 GeV, and in Fig. 6 we took 200 GeV, leading typically to light Higgsinos and winos.

In conclusion, we have explored how 5D extension of the MSSM may generate large A_t to achieve the observed Higgs mass and have sub-TeV stops, perhaps observable at the LHC. We computed the full one-loop RGEs for all supersymmetric and soft breaking parameters. We find that the magnitude of A_t follows closely that of the magnitude of M_3 and increases as the compactification scale decreases.

Acknowledgments

We would like to thank our collaborators Aldo Deandrea and Moritz McGarrie for their helpful discussions. This work is supported by the National Research Foundation (South Africa).

References

[1] G. Aad *et al.* [ATLAS collaboration], Phys. Lett. B716 (2012) 1 [arXiv:1207.7214].
[2] S. Chatrchyan *et al.* [CMS collaboration], Phys. Lett. B716 (2012) 30 [arXiv:1207.7235].
[3] A. Bharucha, A. Goudelis and M. McGarrie, Eur. Phys. J. C **74**, 2858 (2014) [arXiv:1310.4500 [hep-ph]].
[4] F. Brummer, M. McGarrie and A. Weiler, JHEP **1404**, 078 (2014) [arXiv:1312.0935 [hep-ph]].
[5] S. Abel and M. McGarrie, JHEP **1407**, 145 (2014) [arXiv:1404.1318 [hep-ph]].
[6] G. Bhattacharyya and T. S. Ray, JHEP **1005**, 040 (2010) [arXiv:1003.1276 [hep-ph]].
[7] M. McGarrie, arXiv:1109.6245 [hep-ph].
[8] M. McGarrie, JHEP **1304**, 161 (2013) [arXiv:1303.4534 [hep-th]].
[9] K. R. Dienes, E. Dudas and T. Gherghetta, Phys. Lett. B **436**, 55 (1998) [hep-ph/9803466].
[10] K. R. Dienes, E. Dudas and T. Gherghetta, Nucl. Phys. B **537**, 47 (1999) [hep-ph/9806292].
[11] A. Pomarol, Phys. Rev. Lett. **85**, 4004 (2000) [hep-ph/0005293].
[12] Y. Nomura and D. Poland, Phys. Rev. D **75**, 015005 (2007) [hep-ph/0608253].
[13] A. Abdalgabar and A. S. Cornell, J. Phys. Conf. Ser. **455**, 012050 (2013) [arXiv:1305.3729 [hep-ph]].
[14] A. Abdalgabar, A. S. Cornell, A. Deandrea and M. McGarrie, JHEP **1407**, 158 (2014) [arXiv:1405.1038 [hep-ph]].
[15] M. S. Carena, J. R. Espinosa, M. Quiros and C. E. M. Wagner, Phys. Lett. B **355**, 209 (1995) [hep-ph/9504316].
[16] H. E. Haber, R. Hempfling and A. H. Hoang, Z. Phys. C **75**, 539 (1997) [hep-ph/9609331].
[17] C. A. Scrucca, Mod. Phys. Lett. A **20**, 297 (2005) [hep-th/0412237].
[18] A. Falkowski, H. M. Lee and C. Ludeling, JHEP **0510**, 090 (2005) [hep-th/0504091].
[19] A. Delgado, hep-ph/0409073.

High Energy Particle Physics Workshop (HEPPW2015) IOP Publishing
Journal of Physics: Conference Series **645** (2015) 012002 doi:10.1088/1742-6596/645/1/012002

Some theories beyond the Standard Model

Alan S. Cornell

National Institute for Theoretical Physics, School of Physics and Mandelstam Institute for Theoretical Physics, University of the Witwatersrand, Johannesburg, Wits 2050, South Africa

E-mail: alan.cornell@wits.ac.za

February 2015

Abstract. A brief review on the physics beyond the Standard Model is given, as was presented in the High Energy Particle Physics workshop on the 12th of February 2015 at the iThemba North Labs. Particular emphasis is given to the Minimal Supersymmetric Standard Model, with mention of extra-dimensional theories also.

1. Introduction

The Standard Model (SM) of elementary particle physics, developed in the 1970's, describes the behaviour of all known elementary particles with impressive accuracy, and it all traces back to the relatively simple mathematical laws within the formalism of quantum field theory. As such, symmetry groups play a particularly important role in this context, where our three generations of quarks and leptons have their interactions mediated by the gauge bosons. However, the most mysterious field within the SM is the Higgs field, which forms a condensate filling the whole Universe. The motion of all particles is influenced by this condensate, which is how the particles gain their mass. Recall that the Higgs boson, a spin-0 particle produced from excitations of the Higgs field, was recently discovered, completing the particle spectrum of the SM, but that doesn't mean the story is now over. The study of the Higgs boson's couplings is now extremely important, and is connected to beyond the SM (BSM) physics.

Figure 1. *The one-loop contributions to the Higgs mass, where the left diagram contains the fermion loops, the middle contributions are from the gauge bosons, and the right from Higgs self-interactions.*

The structure of the radiative corrections to the Higgs boson is quite different from those of fermions and gauge bosons. To one-loop the renormalised Higgs mass is:

$$m_h^2 = m_{h0}^2 - \frac{3}{8\pi^2} y_t^2 \Lambda^2 + \frac{1}{16\pi^2} g^2 \Lambda^2 + \frac{1}{16\pi^2} \lambda^2 \Lambda^2 \,, \tag{1}$$

where Λ is the scale at which the loop integrals are cut off. As such, the corrections to the Higgs mass, which are referred to as "quadratic divergences", are related to Λ. If Λ is high, the correction is extremely large with respect to the on-shell Higgs mass. This is the "fine-tuning problem", where such quadratic divergences are not radiative corrections to fermion or gauge boson masses (these being protected by chiral and gauge symmetries).

If there is an intermediate scale at which new physics manifests, this problem is resolved and the radiative corrections from any new particles ameliorates issues arising from the SM. Note that merely adding new particles that couple to the Higgs boson, to cancel one loop level quadratic corrections, is insufficient as such chance cancellations do not guarantee cancellations to all orders in a perturbative theory. To guarantee cancellations of these quadratic divergences we need a new symmetry. Some of the established approaches to handle this, which shall be reviewed here are:

1. **Supersymmetry** (SUSY): a symmetry between bosons and fermions, where the chiral symmetry of fermions controls the Higgs mass divergences by cancelling the diagrams containing SM fields with diagrams featuring their sparticles (their supersymmetric partners).
2. **Extra-dimensions**: whilst we seem to reside in a four-dimensional spacetime, we may actually reside in five or more spactime dimensions. The additional or extra-dimensions are somehow "inaccessible" to us (perhaps being compactified in some way), meaning that the true Planck mass could be of comparable scale to the electroweak scale.

Current experimental measurements greatly constrain each of these models, where the SM remains our most successful model to date. However, we may find deviations in the future if we focus on the few fundamental parameters present in the SM, such as the gauge couplings and the vacuum expectation value (vev) of the Higgs; as the measurements of these parameters becomes more precise, greater constraints can be placed on any BSM theory.

Returning to the quadratic divergences above, the divergence in Eq. (1) is estimated by Λ, a cut-off we introduced to the loop integrals. As this momentum cut-off is independent of the external momenta, the divergences are regularisation dependent objects. For dimensional regularisation these divergences are then trivially zero. As such is the fine-tuning problem something which should be taken seriously?

Table 1. The additional particle content of the MSSM.

Names	Spin	P_R	Gauge Eigenstates	Mass Eigenstates
Higgs bosons	0	+1	$H_u^0\ H_d^0\ H_u^+\ H_d^-$	$h^0\ H^0\ A^0\ H^\pm$
squarks	0	-1	$\tilde{u}_L\ \tilde{u}_R\ \tilde{d}_L\ \tilde{d}_R$	(same)
			$\tilde{s}_L\ \tilde{s}_R\ \tilde{c}_L\ \tilde{c}_R$	(same)
			$\tilde{t}_L\ \tilde{t}_R\ \tilde{b}_L\ \tilde{b}_R$	$\tilde{t}_1\ \tilde{t}_2\ \tilde{b}_1\ \tilde{b}_2$
sleptons	0	-1	$\tilde{e}_L\ \tilde{e}_R\ \tilde{\nu}_e$	(same)
			$\tilde{\mu}_L\ \tilde{\mu}_R\ \tilde{\nu}_\mu$	(same)
			$\tilde{\tau}_L\ \tilde{\tau}_R\ \tilde{\nu}_\tau$	$\tilde{\tau}_1\ \tilde{\tau}_2\ \tilde{\nu}_\tau$
neutralinos	1/2	-1	$\tilde{B}^0\ \tilde{W}^0\ \tilde{H}_u^0\ \tilde{H}_d^0$	$\tilde{N}_1\ \tilde{N}_2\ \tilde{N}_3\ \tilde{N}_4$
charginos	1/2	-1	$\tilde{W}^\pm\ \tilde{H}_u^+\ \tilde{H}_d^-$	$\tilde{C}_1^\pm\ \tilde{C}_2^\pm$
gluino	1/2	-1	$\tilde{g}$	(same)
goldstino	1/2	-1		(same)
(gravitino)	3/2	-1	$\tilde{G}$	(same)

2. Supersymmetry

This fine-tuning problem, as presented above with a momentum cut-off, is valid when at some higher scale a theory with a larger symmetry exists. For the case of SUSY models our regularisation scheme must satisfy our SUSY and we cannot remove all quadratic divergences. Therefore, the radiative corrections to the Higgs sector are proportional to the mass scale of the sparticles (the SUSY scale) under a proper regularisation. Taking the sparticle masses as being much greater than the SM particle masses, the theory appears as the SM at low energies with Λ being at the SUSY scale. This is an example where fine-tuning arguments hold for theories with an extended scale above which a new symmetry arises.

SUSY is a symmetry which exchanges fermions for bosons, and vice-versa. The Minimal Supersymmetric SM (MSSM) is where we extend the SM by having a SUSY in the limit when particle masses are negligible, and where this is presumed to be an effective theory of some fully supersymmetric model. As the full theory undergoes a spontaneous SUSY breaking, the sparticles obtain a mass that is much greater than those for their partner SM particles. The fermion's superpartner is a spin-0 particle (a sfermion), whilst a gauge boson's superpartner has spin-1/2 (a gaugino). For the Higgs boson, its superpartner has spin-1/2 also (a higgsino). In Tab. 1 the particle content of the MSSM is listed, where the superpartners have the same charges as their SM counter-parts. This is due to the generators of the SUSY transform commuting with the $SU(3) \times SU(2) \times U(1)$ transformation of the SM. Note also that in the MSSM we have two Higgs doublets and two higgsinos, as chiral fermions with charge $(1,2)_{\pm 1/2}$ are constrained by anomaly cancellation.

From Tab. 1 we can see that the particle content has doubled from the SM. However, the SUSY does not determine the masses of the sparticles, even though all dimensionless couplings with these particles (such as Yukawa and four point couplings) are. The relationships between couplings can be understood only from the full supersymmetric

Journal of Physics: Conference Series **645** (2015) 012002 doi:10.1088/1742-6596/645/1/012002

theory.

Let us consider some features of supersymmetric models:

- As the quadratic divergence arising from the top loop is cancelled by the stop loop (from a Higgs-Higgs-stop-stop interaction), given that both diagrams are proportional to y_t^2, etc., there are no quadratic divergences in this theory. However, as scalar particles are in the same multiplet as fermions (fermion mass being logarithmically divergent), the Higgs quartic coupling is proportional to the square of the higgsino-gaugino loops. Therefore fine-tuning within the Higgs sector is greatly reduced.
- As the four point coupling of the Higgs is now a gauge coupling, it is always positive at the Planck scale. Please see the work, including proceedings, of Abdalgabar *et al.* [1] and references therein.
- From gauge invariance there are no baryon and lepton number violating processes in the SM. This is not the case in SUSY models as higgsinos carry the same quantum numbers as leptons.
- Gauge coupling: In models such as the MSSM the number of particles can be doubled from the SM, with runnings of the gauge couplings being modified above the sparticle mass scale. As such the gauge couplings unify much better than in the SM case at the GUT scale (see Fig. 2). This means that a supersymmetric GUT agrees with current experimental results, even though fine-tuning issues may persist, as the Higgs sector may violate the GUT symmetry.

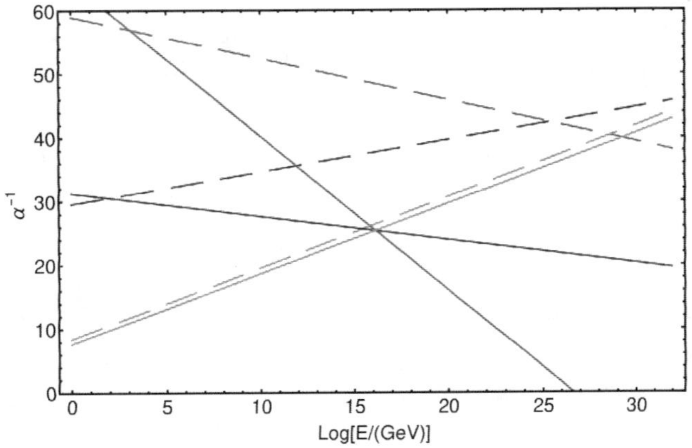

Figure 2. *The one loop renormalisation group evolution of the gauge couplings in the SM (dashed lines) and MSSM (solid lines).*

2.1. Origins of SUSY breaking

As a final discussion point on SUSY, models like the MSSM are incomplete theories as the necessary SUSY breaking mechanism comes from elsewhere. Note the SUSY breaking set-up has, in general, some hidden sector containing fields which spontaneously break the SUSY. The hidden sector does not couple directly to the visible sector but does so through some messenger sector, where the messenger particles have some mass scale.

However, this messenger sector is already severely constrained for the MSSM sector from such measurements as flavour changing neutral currents ($K^0 - \bar{K}^0$ mixing for example). Even so, obtaining information on the hidden sector remains difficult. In some cases the gravitino (the graviton's superpartner) mass $m_{3/2} = F_0/M_{Pl}$, where F_0 is the total energy of SUSY breaking, could be the lightest supersymmetric partner (LSP). Often the next to LSP (NLSP) would be long-lived in such cases. As such the NLSP may be detectable at colliders, with its life-time giving information on this hidden sector responsible for SUSY breaking. Note that when the gravitino is not the LSP, the gravitino will be long enough lived to effect the big bang nucleosynthesis.

The mechanisms of the messenger sector set the scale of the sparticle masses, where the on-shell SUSY particle masses are then determined from the renormalisation group equations being run down to the lower energy scale. A more complete discussion can be found in Ref. [2].

3. Extra-dimensions

Among the other possible models which address the fine-tuning problem (we shall review) are those with extra-dimensions. These models address this problem by noticing that the observed Planck scale is effective and that the true (higher-dimensional) Planck scale can be smaller and even of the order of the electroweak scale. As such the parameters in the Higgs sector are of the order of the true Planck scale and not the effective one. Furthermore, in some extra-dimensional models the Higgs may be the fifth-dimensional component of the gauge field, and as such its parameters are protected by the gauge symmetry (no quadratic divergences).

To given an overview of some of the ideas used in these models, let us consider the case where we have additional spatial dimensions that are compactified with radius R [3]. If we have one flat extra-dimension, then fields propagating in this extra-dimension must obey a periodic boundary condition, for example

$$\phi(x,y) = \phi(x,y+R)\,, \tag{2}$$

with x being our usual four spacetime dimensions, and y our extra-dimension.

With such a boundary condition wavefunctions can be written as

$$\psi(x, y) = \psi'(x)\exp(ip_5 y) , \tag{3}$$

where p_5 is the fifth component of our 5-momenta and satisfies $p_5 R = 2\pi n$ for an integer n. The equation of motion for a particle moving in the additional dimension becomes

$$E_n^2 = p^2 + p_5^2 = p^2 + (2\pi)^2 \left(\frac{n}{R}\right)^2 . \tag{4}$$

That is, we have an infinite tower of massive particles in the effective four-dimensional theory, with masses equal to the discrete values of p_5^2.

The couplings in the higher-dimensional theory are related to those in the effective four-dimensional case, however, this can be non-trivial, depending on the model. For gauge couplings in our simple case above

$$\int d^4x dx_5 \frac{1}{g_5^2} F_{\mu\nu}F^{\mu\nu} \to \int d^4x \frac{1}{g_4^2} F_{\mu\nu}F^{\mu\nu} , \tag{5}$$

where $g_4 = g_5/\sqrt{R}$. So as the extra-dimension becomes larger, g_4 is reduced. This is also true for gravitational interactions. The four-dimensional gravitational interaction may be very weak if the size of the extra-dimensions are very large. In the case of large extra-dimensional models these can solve the fine-tuning problem as discussed above, by making the true Planck scale in the higher-dimensional theory much smaller.

There are many other varieties of extra-dimensional models, and they are not all flat. A famous example is that of the Randall-Sundrum model [4] where the additional spatial dimension has the "warped" metric

$$ds^2 = e^{-2\sigma(\phi)}\eta_{\mu\nu}dx^\mu dx^\nu + r_c^2 d\phi^2 , \tag{6}$$

and where the boundaries of our additional dimension are $\phi = 0$ and π. In this case the action becomes

$$S_{gravity} = \int d^4x \int_{-\pi}^{+\pi} d\phi\sqrt{-G} - \Lambda + 2M^3 R , \tag{7}$$

with "warp" factor

$$\sigma(\phi) = r_c|\phi|\sqrt{\frac{-\Lambda}{24M^3}} , \tag{8}$$

together with appropriate actions on the boundaries.

4. Concluding remarks

As a final thought, note that there are other indications of the existence of new physics between the weak scale and the Planck scale. Firstly, consider the Higgs potential and how it evolves with energy in the SM, in particular its stability. The potential is a function of the top and Higgs masses, and current top and Higgs mass measurements favour a metastable Higgs potential [5]. Now, there is no reason that the Higgs vev should fall in such a metastable region, and this also suggests that additional particles that couple to the Higgs sector change the shape of the potential dramatically, see

Journal of Physics: Conference Series **645** (2015) 012002 doi:10.1088/1742-6596/645/1/012002

Refs. [6] for example. So further analysis of this region of parameter space may give indications of BSM physics.

Though we have not discussed it in this review, we note that another topical result which requires some new physics is dark matter. From various cosmological and astrophysical observations we know that $\sim 27\%$ of the energy content of the Universe is in the form of dark matter. And therefore such particles must be stable and neutral. Many candidates for dark matter particles exist in a range of BSM theories, including the MSSM, but more studies are required and this is an ongoing field of research.

The points that have been raised in this brief proceedings will hopefully motivate the participants of this workshop to further readings on these and related topics, such as Refs. [2, 7, 8, 9].

References

[1] A. Abdalgabar, A. S. Cornell, A. Deandrea and M. McGarrie, JHEP **1407**, 158 (2014) [arXiv:1405.1038 [hep-ph]].

[2] L. J. Hall, J. D. Lykken and S. Weinberg, Phys. Rev. D **27**, 2359 (1983).

[3] I. Antoniadis, N. Arkani-Hamed, S. Dimopoulos and G. R. Dvali, Phys. Lett. B **436**, 257 (1998) [hep-ph/9804398]; N. Arkani-Hamed, S. Dimopoulos and G. R. Dvali, Phys. Lett. B **429**, 263 (1998) [hep-ph/9803315]; N. Arkani-Hamed, S. Dimopoulos and G. R. Dvali, Phys. Rev. D **59**, 086004 (1999) [hep-ph/9807344].

[4] L. Randall and R. Sundrum, Phys. Rev. Lett. **83**, 3370 (1999) [hep-ph/9905221]; L. Randall and R. Sundrum, Phys. Rev. Lett. **83**, 4690 (1999) [hep-th/9906064].

[5] S. Alekhin, A. Djouadi and S. Moch, Phys. Lett. B **716**, 214 (2012) [arXiv:1207.0980 [hep-ph]].

[6] L. X. Liu and A. S. Cornell, Phys. Rev. D **86**, 056002 (2012) [arXiv:1204.0532 [hep-ph]]; A. Abdalgabar, A. S. Cornell, A. Deandrea and A. Tarhini, Eur. Phys. J. C **74**, no. 5, 2893 (2014) [arXiv:1307.6401 [hep-ph]].

[7] R. Rattazzi, *Cargese 2003, Particle physics and cosmology* 461-517 [hep-ph/0607055].

[8] F. Arneodo, arXiv:1301.0441 [astro-ph.IM].

[9] C. Strege, G. Bertone, F. Feroz, M. Fornasa, R. Ruiz de Austri and R. Trotta, JCAP **1304**, 013 (2013) [arXiv:1212.2636 [hep-ph]].

Quasi-normal Modes for Spin-3/2 Fields

Gerhard Erwin Harmsen

National Institute for Theoretical Physics, School of Physics and Mandelstam Institute for Theoretical Physics, University of the Witwatersrand, Johannesburg, Wits 2050, South Africa

E-mail: gerhard.harmsen5@gmail.com

Abstract. With the Large Hadron Collider already able to produce collisions with an energy of 8 TeV, the formation of higher dimensional black holes may soon be possible. In order to determine if we are detecting these higher dimensional black holes we need to have a theoretical understanding of what the signatures of such black holes could be. As such we shall discuss quasi-normal modes (QNMs) for spin-3/2 fields as they travel through a black hole background. We will begin by studying possible QNMs for N-dimensional Schwarzschild black holes and extend by looking at N-dimensional Kerr black holes. We will use the Wentzel-Kramers-Brillouin approximation to determine the QNMs for the two types of black holes described above.

1. Introduction

There are currently many ways to indirectly detect black holes [1]. One possible direct way of detecting a black hole is through gravitational wave detection. These are oscillations in the space time and are emitted by black holes when they are perturbed through either black hole-black hole collisions, stellar collapse or other matter interactions. These perturbations cause non-radial oscillations on the surface of a black hole and are called quasi-normal modes (QNMs). These QNMs in turn effect the space-time around a black hole and result in quasi-normal frequencies being emitted from black holes, these frequencies would be detected as gravitational waves, or particles [2]. What is interesting is that the allowed quasi-normal frequencies of a black hole are directly related to the characteristics of the black holes. For instance, the allowed quasi-normal frequencies of a Schwarzschild black hole are given by the mass of the black hole, but for a Kerr black hole they are determined by the angular momentum, charge and mass of the black hole [2]. This means that we could not only directly detect black holes by detecting their gravitational emission, but we could also get an idea of the parameters of the black hole by using this. A more elaborate discussion on detecting these QNMs is given in Ref. [3]. It has been shown that the allowed QNMs of a black hole can be calculated using perturbative methods such as the Wentzel-Kramers-Brillouin (WKB) method [4] and the improved asymptotic iteration method (AIM) [5]. Where the validity for determining QNMs using these methods has been shown. We will use these methods to determine the allowed QNMs for spin-3/2 fields. We shall first give a brief overview of black holes and the metrics that we will be using, in order to describe the space time surrounding these black holes. We then give a brief discussion of QNMs and then give an overview of the WKB method and improved AIM.

Journal of Physics: Conference Series **645** (2015) 012003 doi:10.1088/1742-6596/645/1/012003

2. Black Holes

Black holes have been described as the hydrogen atom of general relativity, since they are relatively simple general relativistic objects which display all of the properties predicted by Einstein's field equations. Hence a better understanding of black holes would allow us to better understand general relativity, just as the hydrogen atom can help in our understanding of quantum mechanics. In this paper we will set c = 1.

2.1. Schwarzchild Black holes

The Schwarzschild metric can be used to describe the space time surrounding any non-rotating objects in a vacuum [6]. It is given as:

$$ds^2 = -\left(1 - \frac{2GM}{r}\right)dt^2 + \left(1 - \frac{2GM}{r}\right)^{-1} dx^2 + r^2\left(d\phi^2 + \sin(\theta)d\theta^2\right), \tag{1}$$

where $G = 7.426 \times 10^{-28} m/kg$, the mass ($M$) denotes the mass of the object and r denotes the radial distance between a point in the space time and the center of that object. This metric is valid for all values of r greater than zero. We can describe the space time surrounding a black hole by looking at systems for which the value of $2GM$ is large compared to r. When r approaches $2GM$ our metric will produce a singularity, where this singularity is in fact a "coordinate singularity", which means that there exists a problem with the coordinate system we have chosen rather than there being any actual physical singularity at that point in space-time [6]. The point $r = 2GM$ is in fact the radial location of the event horizon of the black hole. These types of black holes are the simplest types of black holes and provide the simplest case for studying QNMs.

2.2. Kerr black holes

Although the Schwarzschild black holes are simple non-trivial examples of black holes, we are more interested in studying possible QNMs for a spinning black hole; as these are more likely to form as it is unlikely that matter would collapse with zero angular momentum. As before, we will use a metric to describe the space-time surrounding a black hole [7]. Where the metric we will use is the Kerr metric, given as:

$$\begin{aligned} ds^2 = &-\left(1 - \frac{2GMr}{r^2 + a^2\cos^2(\theta))}\right)dt^2 + \left(\frac{r^2 + a^2\cos^2\theta}{r^2 - 2GMr + a^2}\right)dr^2 + \left(r^2 + a^2\cos^2\theta\right)d\theta^2 \\ &+ \left(r^2 + a^2 + \frac{2GMra^2\sin^2\theta}{r^2 + a^2\cos^2\theta}\right)\sin^2\theta d\phi^2 - \left(\frac{4GMra\sin^2\theta}{r^2 + a^2\cos^2\theta}\right)d\phi dt, \end{aligned} \tag{2}$$

where G, M and r represent the same quantities as they did for the Schwarzchild metric. The new term in this metric is the "a" term, which describes the rate of rotation for the black hole. If we take this "a" value to zero we see that the "$d\phi dt$" term goes to zero. So through a simple analysis we can see that this metric describes a black hole rotating in the ϕ plane.

3. Quasi-Normal Modes

As stated before, QNMs are the reaction to a black hole being perturbed in some way. A simple explanation of what a QNM is can be described by the tapping of a wine glass. If we tap the wine glass we are in fact perturbing the glass. This perturbation causes the glass to resonate at specific frequencies. The allowed frequencies are determined by the composition of the glass, and even the contents within the glass. If the glass continued to resonate forever then we would be generating a normal mode. This would be an example of the standing wave, as many have studied in first year physics courses. Since the glass does stop resonating we say it is a QNM. So when studying QNMs we are studying standing waves with some damping term attached. If we consider our wine glass example we should note that wine glasses with different amounts of water in them ring at different frequencies, so we could, in theory, determine the amount of water in a glass by the sound it makes when we tap the glass. A similar idea can be used for the emitted quasi-normal frequencies of a black hole. Although with a black hole energy is being radiated, not a sound, but rather as disruptions in the space time which we call gravitational waves.

3.1. Tortoise coordinate

It is common to use tortoise coordinates when studying black holes, where these are defined as follows [8]:

$$r* = r + 2GM \ln\left(\frac{r}{2GM} - 1\right). \tag{3}$$

This then satisfies the following relation:

$$\frac{dr*}{dr} = \left(1 - \frac{2GM}{r}\right)^{-1}. \tag{4}$$

The consequence of using this coordinate system is that as r tends to $2GM$, $r*$ goes to negative infinity, and as r tends to infinity, $r*$ tends to positive infinity. These two points correspond to the event horizon and a point infinitely far from the black hole respectively. We will use tortoise coordinates in order to develop a mathematical description of QNMs in a black hole background.

3.2. A Mathematical description of QNMs

In order to develop a more mathematical understanding of QNMs we can look at the mathematical formula for standing waves since, as stated above, QNMs are merely damped standing waves (mathematically speaking). So we can represent these QNMs as follows [9]:

$$\frac{d^2}{dx^2}\Psi - \frac{d^2}{dt^2}\Psi - V\Psi = 0. \tag{5}$$

With Ψ describing some wavelike function where x and t denote space and time coordinates respectively. V is some x-dependent potential. We can solve the equation as we would a standing wave problem and so we can assume the following time dependency [3]:

$$\Psi(x,t) = e^{-i\omega t}\phi(x). \tag{6}$$

Plugging this into our equation for a QNM we get:

$$\frac{d^2\phi(x)}{dx^2} - (\omega^2 + V(x))\phi(x) = 0. \tag{7}$$

This is the general form for the QNMs we will be studying in this paper. In order to solve this equation we require specific boundary terms, similar to those required for solving the standing wave equation. The boundary conditions that we will impose are:

$$\begin{aligned} V &\to 0; x \to \infty, \\ V &\to 0; x \to -\infty. \end{aligned} \tag{8}$$

This means that particles fall into the black hole at the horizon, and those which move to infinity are no longer influenced by the black hole. In order to determine if a frequency of a field is in fact a QNM, we need to ensure that it obeys these requirements. In this project we will use both the WKB method, to 6th order [4] , and the improved AIM [5] to determine the solutions to our QNM equation.

4. Aproximation Methods

4.1. WKB Approximation

The WKB method is a well known approximation, and is usually used in quantum mechanics to determine the solutions of Schrödinger equations in systems with non-constant potentials. The WKB method can be used to determine the approximate solution to second order differential equations. A brief example of how the method is used to solve second order differential equations is given below.

We can use the WKB approximation to solve problems of the following form [10]:

$$\epsilon^2 \frac{d^2 y}{dx^2} = Q(x)y, \tag{9}$$

where $\epsilon \ll 1$ and positive. We can solve this equation by assuming the following solution:

$$y(x,\epsilon) = A(x,\epsilon)e^{\frac{iu(x)}{\epsilon}}. \tag{10}$$

Taking the first and second derivatives of this function we have that:

$$y(x,\epsilon) = A(x,\epsilon)e^{\frac{iu(x)}{\epsilon}},$$

$$y'(x,\epsilon) = A'(x,\epsilon)e^{\frac{iu(x)}{\epsilon}} + A(x,\epsilon)\left(\frac{iu'(x)}{\epsilon}\right)e^{\frac{iu(x)}{\epsilon}},$$

$$\begin{aligned} y''(x,\epsilon) &= A''(x,\epsilon)e^{\frac{iu(x)}{\epsilon}} + A'(x,\epsilon)\left(\frac{iu'(x)}{\epsilon}\right)e^{\frac{iu(x)}{\epsilon}} + A'(x,\epsilon)\left(\frac{iu'(x)}{\epsilon}\right)e^{\frac{iu(x)}{\epsilon}} - A(x,\epsilon)\left(\frac{u'(x)}{\epsilon}\right)^2 e^{\frac{iu(x)}{\epsilon}} \\ &= \left(A'' + 2A'\frac{iu'(x)}{\epsilon} + A(\frac{iu''(x)}{\epsilon} - \left(\frac{u'(x)}{\epsilon}\right)^2\right)e^{\frac{iu(x)}{\epsilon}}. \end{aligned} \tag{11}$$

Plugging the relevant equations Eq.(11) into Eq.(9) we get:

$$\epsilon^2\left(A'' + 2A'\frac{iu'(x)}{\epsilon} + A(\frac{iu''(x)}{\epsilon} - \left(\frac{u'(x)}{\epsilon}\right)^2\right) - Q(x)A(x,\epsilon) = 0. \tag{12}$$

We now perform a series expansion on A as follows:

$$\begin{aligned} A(x,\epsilon) &= A_0(x) + \epsilon A_1(x) + \epsilon^2 A_2(x) + ... \\ A'(x,\epsilon) &= A_0'(x) + \epsilon A_1'(x) + \epsilon^2 A_2'(x) + ... \\ A''(x,\epsilon) &= A_0''(x) + \epsilon A_1''(x) + \epsilon^2 A_2''(x) + ... \end{aligned} \tag{13}$$

so that Eq.(12) becomes

$$\begin{aligned} &\epsilon^2\left(\left(A_0''(x) + \epsilon A_1''(x) + \epsilon^2 A_2''(x) + ...\right) + 2\left(A_0'(x) + \epsilon A_1'(x) + \epsilon^2 A_2'(x) + ...\right)\frac{iu'(x)}{\epsilon}\right) + \\ &\epsilon^2\left(\left(A_0(x) + \epsilon A_1(x) + \epsilon^2 A_2(x) + ...\right)(\frac{iu''(x)}{\epsilon} - \left(\frac{u'(x)}{\epsilon}\right)^2 - Q(x)\left(A_0(x) + \epsilon A_1(x) + \epsilon^2 A_2(x) + ...\right)\right) = 0. \end{aligned} \tag{14}$$

We then group the terms in powers of ϵ to get our approximation. So for instance we can collect all terms of ϵ^0 to get a solution to $u(x)$

$$\begin{aligned} A_0(x)u'(x)^2 + A_0 Q(x) &= 0, \\ u'(x) &= \pm\sqrt{Q(x)}, \\ u(x) &= \pm\int_{x_0}^{x} \sqrt{Q(k)}dk. \end{aligned} \tag{15}$$

We then take orders of ϵ to get our first order approximation of A

$$\begin{aligned} -u'(x)^2 A_1 + iA_0 u''(x) + 2iA_0' u'(x) + Q(x)A_0 &= 0, \\ A_0 u''(x) + 2A_0' u'(x) &= 0, \\ \frac{A_0}{2\sqrt{Q(x)}} + \frac{A_0'}{\sqrt{Q(x)}} &= 0, \end{aligned} \tag{16}$$

$$\int \frac{1}{A_0} dA_0 = \int \frac{1}{2Q(x)} dx. \tag{17}$$

We can use the WKB method in order to study black hole perturbations since these equations reduce to the form of a wave equation, similar to the example above [3]. This equation, with the known potential, is then solved using the 6th order WKB approximation. The cases for Schwarzschild black holes and Reissner-Nordström black holes has been studied extensively [5], these studies have not considered the case of spin-3/2 fields. The Kerr black hole requires us to solve a coupled wave equation for the radial and the angular case, hence this situation has not been studied as extensively as the other cases.

4.2. Improved AIM

The theory of this method is given in Ref. [11]. The theorem of this method is given below: Given λ_0 and s_0 in $C_\infty(a,b)$, then the differential

$$y'' = \lambda_0(x)y' + s_0(x)y, \tag{18}$$

has the general solution of:

$$y(x) = exp\left(-\int^x \alpha dt\right)\left[C_2 + C_1\int^x exp\left(\int^t (\lambda_0(\tau) + 2\alpha(\tau))\, d\tau\right)\right], \tag{19}$$

for some $n > 0$.

$$\frac{s_n}{\lambda_n} = \frac{s_{n-1}}{\lambda_{n-1}} \equiv \alpha, \tag{20}$$

where $\lambda_k = \lambda_{k-1}' + s_{k-1} + \lambda_0\lambda_{k-1}$ and $s_k = s_{k-1}' + s_0\lambda_{k-1}$ for $k = 1, 2, ..., n$.
We again solve for the QNMs by first obtaining the potential energy term for a spin-3/2 particle in a black hole background, and then use the AIM approximation to obtain the allowed QNMs.

5. Concluding remarks

We are currently checking the results obtained in Ref. [12] using the WKB method, and shall expand on this work by checking these results against those obtained by using the improved AIM Ref. [5]. We can then extend this by considering potentials for a spin-3/2 particle in a Kerr black hole background. We would like to, with time permitting, calculate the allowed QNMs for a spin-3/2 particle in a Reissner-Nordström black hole, which is a stationary charged black hole.

Journal of Physics: Conference Series **645** (2015) 012003 doi:10.1088/1742-6596/645/1/012003

Acknowledgments

I would like to thank Prof. Alan Cornell and Prof. Hing-Tong Cho for their imparting some of their knowledge to me on this topic. I would also like to acknowledge the NRF and NITheP for funding me and this research.

References

[1] Novello M, Visser M and Volovik G E 2002 *Artificial black holes* vol 5 (World Scientific)
[2] Leaver E W 1985 *Proceedings of the Royal Society of London. A. Mathematical and Physical Sciences* **402** 285–298
[3] Kokkotas K D and Schmidt B G 1999 *Living Rev. Rel* **2** 262
[4] Konoplya R 2003 *Physical Review D* **68** 024018
[5] Cho H, Cornell A, Doukas J and Naylor W 2010 *Classical and Quantum Gravity* **27** 155004
[6] MooreThomas 2013 *AGeneralRelativityWorkbook* (University Science Books)
[7] Chandrasekhar S 1983 *Research supported by NSF. Oxford/New York, Clarendon Press/Oxford University Press (International Series of Monographs on Physics. Volume 69), 1983, 663 p.* **1**
[8] Gui-hua T, Wang S k and Zhong S 2006
[9] Vishveshwara C 1970
[10] Griffiths D J and Harris E G 1995 *Introduction to quantum mechanics* vol 2 (Prentice Hall New Jersey)
[11] Ciftci H, Hall R L and Saad N 2003 *Journal of Physics A: Mathematical and General* **36** 11807
[12] Piedra O P F 2011 *Int.J.Mod.Phys.* **D20** 93–109 (*Preprint* 1006.3327)

Short Path Length Energy Loss in the Quark-Gluon Plasma from pQCD

Isobel Kolbe[1] **and W. A. Horowitz**[1]

[1]Department of Physics, University of Cape Town, Private Bag X3, Rondebosch 7701, South Africa

E-mail: isobel.kolbe@gmail.com

Abstract. An outline for research to be done is given. In the heavy ion experiments at RHIC and the LHC, it is widely believed that a state of matter known as the quark-gluon plasma (QGP) has been produced. The so-called hard particles, or particles with very high momentum that are produced as a consequence of the asymptotic freedom of QCD, can be used as tomographic probes of the QGP. We will study the way in which energy is dissipated in this QGP by calculating, in pQCD, short path length corrections to the well-known energy loss formulae. This calculation is necessary to address the discovery at the LHC that shockingly small systems appear to exhibit collective behaviour.

1. Introduction

In the heavy ion collisions at both the LHC and RHIC, there is overwhelming evidence that a new state of matter has been created, known as the quark-gluon plasma [1]. Naturally, one would like to study this new phase of matter. However, one of the major challenges such a study is faced with is the extremely short life time of the QGP which excludes the possibility of using an externally produced probe. For this reason, one is reliant on the use of self-generated probes if one is to attempt a characterization of the medium. That is, one must depend on probes that are produced along with the medium.

'Hard partons' or partons (such as a quark or a gluon) with very high transverse momentum can be used as tomographic probes of this new medium in the same way that X-rays are used in medicine, in that one can study the degree to which energy is lost as the parton propagates through the medium [2]. The hard partons that are used in this treatment are a quark anti-quark pair that are produced along with the medium. In rare cases, this pair will be created on the periphery of the medium with the effect that one of the pair will move off away from the medium, passing through numerous stages of thermalization and hadronization before being detected as a high p_T jet in the detector, while the other will have to traverse the medium before being subjected to the same process. This leads to back-to-back jets of which one has a much lower total energy than the other. This phenomenon is known as jet quenching [3].

Jet quenching has been used [4] to explain the discovery [5–8] of high p_T hadron suppression in RHIC experiments. This was done using a pQCD model developed by Gyulassi, Vitev and Lèvai (GLV) which is a *radiative* energy loss model [3].

Several calculations, each improving on or correcting previous works, of the radiative energy loss of a parton moving through the QGP in the GLV model have been done [3, 9–11]. Figure 1

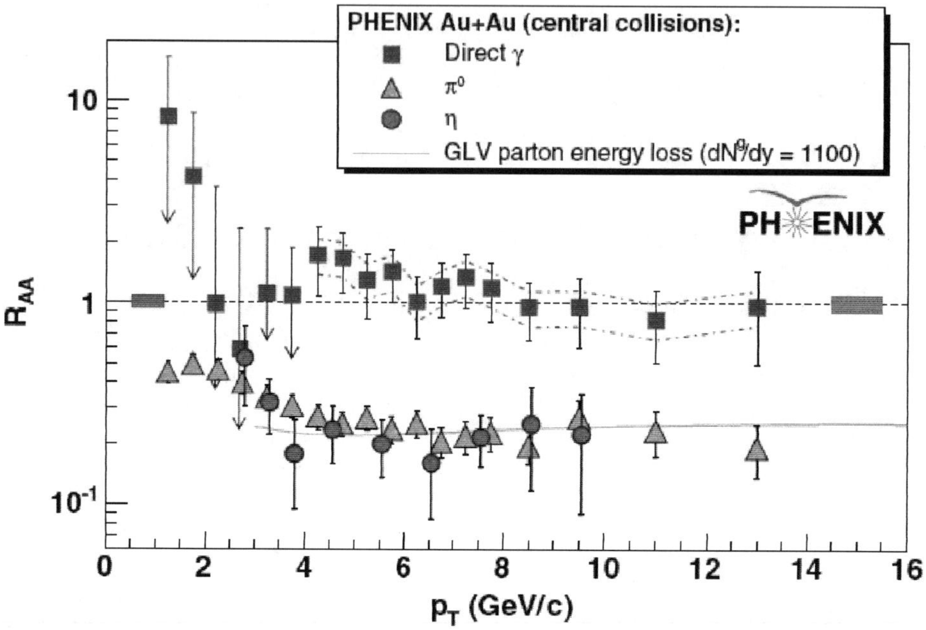

Figure 1. GLV energy loss model prediction compared to RHIC data [12]

shows the considerable success that this model has enjoyed.

It is important to note that Figure 1 also shows that the model controls initial state effects well since we see no suppression of the photon (which is colourless and can therefore move through the medium unperturbed). This is further evidenced in that the suppression of the π^0 and η particles is similar even though their masses differ substantially ($\sim$ 135 MeV and $\sim$ 548 MeV respectively), suggesting that this suppression is a final state effect rather than an initial state effect.

Originally, since this suppression is not seen in smaller systems such as in the $d + Au$ at RHIC [13], it was believed that no medium had been created in these collisions. However, recently, the CMS Collaboration [14] as well as the ALICE Collaboration [15] have released results indicative of collective behaviour in these exceptionally small systems, suggesting that the medium is in fact being created, which is motivation for studying the effects of energy loss in small systems.

The higher luminosities - and therefore higher statistics- of Run II of the LHC will improve the measurements of energy loss. It is therefore highly probable that there will be concrete data for for jet quenching in small systems within the coming years.

We will thus attempt to correct the GLV calculation shown in Figure 1 for short path lengths (such as those present in the very small systems that are displaying collective behaviour).

2. Method and Calculations

The energy loss calculation shown in Figure 1 as the 'GLV parton energy loss' was performed using a number of assumptions:

(i) For the Debye screening length μ_D, the mean free path λ_{MFP} and the size of the system L, the following length scales were assumed

$$\frac{1}{\mu_D} \ll \lambda_{MFP} \ll L. \tag{1}$$

The Debye screening length and mean free path can be calculated from thermal considerations [9] and are therefore known, while the system was considered extremely large since it was on the order of the radius of the heavy ions of the collision.

(ii) The majority of the energy loss is due to radiative loss rather than elastic collisions. This means, amongst others, that only diagrams with up to two or three scattering interactions need to be calculated since very few elastic scatterings occur in the time during which the parton traverses the medium.

This calculation has been done using a reactor operator approach [3, 9] in excess of nine orders of opacity [16], with diagrammatic calculations of upto third order in opacity.

We will relax assumption (i) to account for the smaller system sizes. That is, in our calculation, we will assume rather that

$$\frac{1}{\mu_D} \ll \lambda_{MFP} \sim L \tag{2}$$

This assumption will be used when considering the poles of a number of contour integrals that need to be calculated. Previously, assumption (i) meant that a number of terms pertaining to the potential were neglected since their contribution to the matrix element of the diagram were considered small. We will now keep these terms, meaning that our calculation is largely an addition of terms. It is therefore not necessary to redo the entire calculation.

That said, we have reproduced almost all of the results (that is, we have calculated the majority of the diagrams) obtained by [9].

3. Outlook

As part of a Master's project, we will adjust the assumptions made in the GLV model to compensate for the recent discovery of exceptionally small systems that exhibit collective behaviour. This will entail a number of highly technical calculations but will in essence only involve the retention of previously discarded terms.

References

[1] U. Heinz, "The quark-gluon plasma at RHIC," *Nuclear Physics A*, vol. A721, pp. 30c–39c, 2003.

[2] I. Vitev and M. Gyulassy, "High p_T Tomography of d + Au and Au + Au at SPS, RHIC and LHC," *Phys. Rev. Lett.*, vol. 89, p. 252301, December 2002.

[3] M. Gyulassy, P. Lévai, and I. Vitev, "Reaction Operator approach to Non-abelian Energy loss," *Phys. Rev. Lett.*, vol. 85, p. 5535, 2000.

[4] E. Wang and X. Wang, "Jet tomography of Hot and Cold Nuclear Matter," *Phys. Rev. Lett*, vol. 89, no. 16, p. 162301, 2002.

[5] [PHENIX Collaboration], S. S. Adler, *et al.*, "Suppressed π^0 Production at Large Transverse Momentum in Central Au+Au Collisions at $\sqrt{s_{NN}} = 200$ Gev," *Phys. Rev. Lett.*, vol. 91, p. 072301, 2003.

[6] [PHENIX Collaboration], K. Adcox, *et al.*, "Centrality Dependence of the High p_T Charged Hadron Suppression in Au + Au Collisions at $\sqrt{S_{NN}} = 130$GeV," *Phys. Lett. B*, vol. 561, no. 82, 2003.

[7] [PHENIX Collaboration], K. Adcox, *et al.*, "Suppression of Hadrons with Large Transverse Momentum in Central Au + Au Collisions at $\sqrt{S_{NN}} = 130$GeV," *Phys. Rev. Lett*, vol. 88, p. 022301, 2001.

[8] [STAR Collaboration], C. Adler, *et al.*, "Disappearance of back-to-back high p_T hadron correlations in central Au + Au collisions at at $\sqrt{S_{NN}} = 200$GeV," *Phys. Rev. Lett*, vol. 90, p. 082302, 2003.

[9] M. Djordjevic and M. Gyulassy, "Heavy Quark Radiative Energy Loss in QCD Matter," *Nucl. Phys.*, vol. A733, pp. 265–298, 2004.

[10] M. Gyulassy, P. Lévai, and I. Vitev, "Jet quenching in thin quark-gluon plasmas I: formalism," *Nucl. Phys. B*, vol. 571, pp. 197–233, 2000.

[11] R. Baier, Y. L. Dokshitzer, A. H. Mueller, and D. Schiff, "Medium-induced radiative energy loss; equivalence between the BDMPS and Zakharov formalisms," *Nucl. Phys. B.*, vol. 531(1-3), pp. 403–425, 1998.

[12] J. Lajoie, "Jets and Parton Energy Loss at RHIC: Experimental Status," *Brazilian Journal of Physics*, vol. 37, no. 2C, pp. 825–829, 2007.

Journal of Physics: Conference Series **645** (2015) 012004 doi:10.1088/1742-6596/645/1/012004

[13] [PHENIX Collaboration], S. S. Adler, *et al.*, "Absence of Suppression in Particle Production at Large Transverse Momentum in $\sqrt{s_{NN}} = 200$ Gev $d + Au$ Collisions," *Phys. Rev. Lett.*, vol. 91, p. 072303, 2003.

[14] [CMS Collaboration], "Multiplicity and transverse momentum dependence of two- and four-particle correlations in pPb and PbPb collisions," *Phys. Lett. B*, vol. 724, pp. 213–240, 2013.

[15] [ALICE Collaboration] and J. F. Grosse-Oetringhaus, "Overview of ALICE results at Quark Matter 2014," *Nuclear Physics A*, vol. 931, pp. 22–31, 2014.

[16] S. Wicks, "Up to and beyond ninth order in opacity: Radiative energy loss with GLV," 2008.

[17] [STAR Collaboration], C. Adler, *et al.*, "Centrality Dependence of the High p_T Hadron Suppression in Au + Au Collisions at $\sqrt{S_{NN}} = 130$GeV," *Phys. Rev. Lett*, vol. 89, p. 202301, 2002.

[18] M. Gyulassy, P. Lévai, and I. Vitev, "Jet tomography of AU+Au reactions including multi-gluon fluctuations," *Phys. Lett. B*, vol. 538, p. 282, 2002.

Single Top and Higgs Production in $e^{-}p$ collisions

Mukesh Kumar

National Institute for Theoretical Physics, School of Physics and Mandelstam Institute for Theoretical Physics, University of the Witwatersrand, Johannesburg, Wits 2050, South Africa.

E-mail: mukesh.kumar@cern.ch

Abstract. In this proceedings some studies on the prospects of single top production at the Large Hadron Electron Collider (LHeC) and double Higgs production at the Future Circular Hadron Electron Collider (FCC-he) shall be presented. In particular, we investigated the tbW couplings via single top quark production with the introduction of possible anomalous Lorentz structures, and measured the sensitivity of the Higgs self coupling (λ) through double Higgs production. The studies are performed with 60 GeV electrons colliding with 7 (50) TeV protons for the LHeC (FCC-he).

For the single top studies a parton level study has been performed, and we find the sensitivity of the anomalous coupling at a 95% C.L, considering 10-1% systematic errors. The double Higgs production has been studied with speculated detector parameters and the sensitivity of λ estimated via the cross section study around the Standard Model Higgs self coupling strength (λ_{SM}) considering 5% systematic error in signal and backgrounds. Effects of non-standard CP-even and CP-odd couplings for hhh, hWW and $hhWW$ vertices have been studied and constrained at 95% C.L.

1. Introduction

In the Standard Model (SM) of particle physics the top quark, the heaviest among all matter particles, and the Higgs boson, a particle responsible for giving masses to all matter particles and gauge bosons, play a crucial role in the model. And hence it has been, and still is, a challenge for colliders to study different characteristics of these two particles. These characteristics include their charge, mass, interactions and coupling strengths with other particles etc.

As such we shall briefly review some of the important properties of the top quark and the Higgs boson from theoretical calculations and results from present and past colliders. We will then present some predictions for a future $e^{-}p$ collider known as the Large Hadron Electron Collider (LHeC) and the Future Circular Hadron Electron Collider (FCC-he) at a center of mass energy $\sqrt{s} \approx 1.3$ TeV and $\sqrt{s} \approx 3.5$ TeV, respectively.

The structure of this proceeding shall then be: section 2 is devoted to the top quark studies, sections 3 and 4 are devoted for the Higgs boson studies in $e^{-}p$ collisions. We conclude our inferences based on our studies in section 5.

2. The top quark at the LHeC

In this section we briefly review the physics potential of the proposed LHeC by estimating the accuracy of anomalous Wtb couplings in the single anti-top quark production through $e^{-}p$ collisions [1]. Within the SM the Wtb vertex is purely left-handed. However, the most general

Journal of Physics: Conference Series **645** (2015) 012005 doi:10.1088/1742-6596/645/1/012005

lowest dimension CP conserving (in effect of which the couplings are real) Lagrangian for this vertex is given by

$$\mathcal{L}_{Wtb} = \frac{g}{\sqrt{2}}\left[W_\mu \bar{t}\gamma^\mu \left(V_{tb} f_1^L P_L + f_1^R P_R\right) b - \frac{1}{2m_W} W_{\mu\nu}\bar{t}\sigma^{\mu\nu}\left(f_2^L P_L + f_2^R P_R\right) b\right] + \text{h.c.} \quad (1)$$

Here $f_1^L = 1+\Delta f_1^L$, $W_{\mu\nu} = \partial_\mu W_\nu - \partial_\nu W_\mu$, $P_{L,R} = \frac{1}{2}(1 \mp \gamma_5)$ are left- and right-handed projection operators, $\sigma^{\mu\nu} = \mathrm{i}\left(\gamma^\mu\gamma^\nu - \gamma^\nu\gamma^\mu\right)/2$ and $g = e/\sin\theta_W$. In the SM $|V_{tb}|\, f_1^L \approx 1$. So, along with Δf_1^L all other couplings f_2^L, f_1^R, f_2^R vanish at tree-level, but are non-zero at higher orders. The constraints on these couplings are the following:

- Assuming only one anomalous coupling to be non-zero at a time: $-0.13 \le |V_{tb}| f_1^L \le 0.03$, $-0.0007 \le f_1^R \le 0.0025$, $-0.0015 \le f_2^L \le 0.0004$, $-0.15 \le f_2^R \le 0.57$ from B decays;
- Single top production at DØ assuming $|V_{tb}| f_1^L = 1$: $|f_1^R| \le 0.548$, $|f_2^L| \le 0.224$, $|f_2^R| \le 0.347$;
- Associated tW production at LHC through γp collision: $|f_1^R| \le 0.55$, $|f_2^L| \le 0.22$, $|f_2^R| \le 0.35$;
- ATLAS: asymmetries associated through angular distribution $\mathrm{Re}(f_1^R) \in$ [-0.44,0.48], $\mathrm{Re}(f_2^L) \in$ [-0.24,0.21], $\mathrm{Re}(f_2^R) \in$ [-0.49,0.15].

Loop Corrections:

- QCD: $f_2^R = -6.61 \times 10^{-3}$, $f_2^L = -1.118 \times 10^{-3}$ ($m_t = 171$ GeV);
- EW: $f_2^R = -(1.24 \pm 1.23\mathrm{i}) \times 10^{-3}$, $f_2^L = -(0.102 \pm 0.014\mathrm{i}) \times 10^{-3}$ ($m_h = 126$ GeV);
- SM: $f_2^R = -(7.85 \pm 1.23\mathrm{i}) \times 10^{-3}$, $f_2^L = -(1.220 \pm 0.014\mathrm{i}) \times 10^{-3}$.

2.1. *Single anti-top production*

We analyze the anti-top production through the hadronic and leptonic decay modes of W's as (a) $e^- p \to \bar{t}\nu_e$, $\left(\bar{t} \to W^- \bar{b}, W^- \to jj\right)$ and (b) $e^- p \to \bar{t}\nu_e$, $\left(\bar{t} \to W^- \bar{b}, W^- \to l\nu_l\right)$, respectively. We impose the following selection cuts on events:

- Minimum transverse momenta: $p_{T_{b,j}} \ge 20$ GeV, $p_{T_{b,\bar{l}}} \ge 25$ GeV and $\not{E}_T \ge 25$ GeV;
- Pseudorapidities: $|\eta_{\bar{b},l}| \le 2.5$ and $|\eta_j| \le 2.5$;
- Isolation cuts: $\Delta R_{ij} \ge 0.4$ where i, j are leptons, light-jets and b-jets;
- $\Delta\phi_{\not{E}_T,j} > 0.4$, $\Delta\phi_{\not{E}_T,l} > 0.4$, $\Delta\phi_{\not{E}_T,b} > 0.4$ and
- $|m_{j_1 j_2} - m_W| \le 22$ GeV for the hadronic channel.

After estimation of all possible backgrounds in both channels, imposing the above selection cuts, we observed high yields of single anti-top quark production with fiducial efficiency of ~ 70 % and ~ 90 % in the hadronic and leptonic decay modes of W^- respectively.

2.2. *Estimators and χ^2 analysis*

To find the sensitivity of all non-standard couplings, we follow three different approaches based on one dimensional histograms. In the histograms we compare the SM distributions, including all backgrounds, with all non-standard couplings with representative values $f_1^R = +0.5$, $f_2^L = -0.5$, $f_2^L = +0.5$ and $f_2^R = +0.5$. For hadronic modes, we consider six different distributions, namely, $\Delta\phi_{E_T^{miss} j_1}$, $\Delta\phi_{E_T^{miss} b}$, $\Delta\phi_{E_T^{miss} W}$, $\Delta\phi_{bW}$, $\cos\theta_{bj_1}$ and $\Delta\eta_{bj_1}$, while for leptonic modes, there are four different one dimensional histograms used for the analysis $\Delta\phi_{E_T^{miss} l}$, $\Delta\phi_{E_T^{miss} b}$, $\cos\theta_{bl}$ and $\Delta\eta_{bl}$.

	$A_{\Delta\Phi_{\not E_T j_1}}$	$A_{\Delta\Phi_{\not E_T \bar{b}}}$	$A_{\Delta\Phi_{\not E_T W^-}}$	$A_{\Delta\Phi_{W^- \bar{b}}}$	$A_{\theta_{\bar{b}j_1}}$	$A_{\Delta\eta_{\bar{b}j_1}}$
SM+$\sum_i$ Bkg$_i$	.532 ± .003	.282 ± .005	.503 ± .004	.799 ± .003	.023 ± .001	-.712 ± .003
$f_1^R = +.5$	.327 ± .004	.231 ± .004	.564 ± .004	.778 ± .003	.0005 ± .004	-.806 ± .003
$f_2^L = -.5$	.528 ± .004	.082 ± .004	.716 ± .003	.748 ± .003	-.196 ± .004	-.868 ± .002
$f_2^L = +.5$	.390 ± .005	.269 ± .004	.585 ± .004	.683 ± .004	.106 ± .005	-.795 ± .003
$f_2^R = +.5$	.330 ± .004	.363 ± .004	.566 ± .003	.656 ± .003	-.197 ± .004	-.823 ± .002

Table 1: Asymmetries and errors associated with the kinematic distributions in hadronic histograms at an integrated luminosity $L = 100$ fb^{-1}.

	$A_{\Delta\Phi_{\not E_T l_1}}$	$A_{\Delta\Phi_{\not E_T \bar{b}}}$	$A_{\theta_{\bar{b}l_1}}$	$A_{\Delta\eta_{\bar{b}l_1}}$
SM + $\sum_i$ Bkg$_i$	.384 ± .004	.710 ± .003	.551 ± .006	-.765 ± .007
$f_1^R = +.5$	.484 ± .004	.702 ± .003	.332 ± .006	-.821 ± .003
$f_2^L = -.5$	.526 ± .004	.620 ± .003	.410 ± .006	-.831 ± .002
$f_2^L = +.5$	.353 ± .005	.812 ± .003	.392 ± .007	-.850 ± .003
$f_2^R = +.5$	.424 ± .004	.684 ± .003	.507 ± .005	-.809 ± .003

Table 2: Asymmetries and errors associated with the kinematic distributions in leptonic histograms at an integrated luminosity $L = 100$ fb^{-1}.

2.2.1. Angular Asymmetries from Histograms: As a preliminary study in order to get a feel for different chiral and momentum dependencies of couplings, the following asymmetries are defined:

$$A_{\theta_{ij}} = \frac{N_+^A(\cos\theta_{ij} > 0) - N_-^A(\cos\theta_{ij} < 0)}{N_+^A(\cos\theta_{ij} > 0) + N_-^A(\cos\theta_{ij} < 0)}, \tag{2}$$

$$A_{\Delta\eta_{ij}} = \frac{N_+^A(\Delta\eta_{ij} > 0) - N_-^A(\Delta\eta_{ij} < 0)}{N_+^A(\Delta\eta_{ij} > 0) + N_-^A(\Delta\eta_{ij} < 0)}, \tag{3}$$

$$A_{\Delta\Phi_{ij}} = \frac{N_+^A\left(\Delta\phi_{ij} > \frac{\pi}{2}\right) - N_-^A\left(\Delta\phi_{ij} < \frac{\pi}{2}\right)}{N_+^A\left(\Delta\phi_{ij} > \frac{\pi}{2}\right) + N_-^A\left(\Delta\phi_{ij} < \frac{\pi}{2}\right)}, \tag{4}$$

with $0 \leq \Delta\phi_{ij} \leq \pi$. The asymmetry A_α and its statistical error for N_+^A and N_-^A events, where $N = \left(N_+^A + N_-^A\right) = L \cdot \sigma$, are calculated using the following definition based on binomial distributions:

$$A_\alpha = a \pm \sigma_a, \quad \text{where} \tag{5}$$

$$a = \frac{N_+^A - N_-^A}{N_+^A + N_-^A} \quad \text{and} \quad \sigma_a = \sqrt{\frac{1-a^2}{L \cdot \sigma}}; \quad (\alpha = \cos\theta_{ij}, \Delta\eta_{ij}, \Delta\Phi_{ij}). \tag{6}$$

Here $\sigma \equiv \sigma(e^- p \to \bar{t}\nu, \bar{t} \to W^- \bar{b}) \times BR(W^- \to jj/l^- \bar{\nu}) \times \epsilon_b$ is the total cross section in the respective channels after imposing selection cuts and $\epsilon_b = 0.6$ is the $b/\bar{b}$ tagging efficiency.

Tables 1 and 2 show how asymmetries are affected due to anomalous couplings of the order 10^{-1}. The asymmetry suggests that the distribution of the cosine of the angle between the tagged $\bar{b}$ quark and the highest p_T jet j_1 in the hadronic mode is the most sensitive observable.

2.2.2. Exclusion contour from bin analysis: To make the analysis more effective we perform the χ^2 analysis defined as:

$$\chi^2(f_i, f_j) = \sum_{k=1}^{N} \left(\frac{N_k^{\rm exp} - N_k^{\rm th}(f_i, f_j)}{\delta N_k^{\rm exp}} \right)^2, \tag{7}$$

where $\mathcal{N}_k^{\text{th}}(f_i, f_j)$ and $\mathcal{N}_k^{\text{exp}}$ are the total number of events predicted by the theory involving f_i, f_j and measured in the experiment for the kth bin. $\delta\mathcal{N}_k^{\text{exp}}$ is the combined statistical and systematic error δ_{sys} in measuring the events for the kth bin. If all the coefficient f_i's are small, then the experimental result in the kth bin should be approximated by the SM and background prediction as

$$\mathcal{N}_k^{\text{exp}} \approx \mathcal{N}_k^{\text{SM}} + \sum_i \mathcal{N}_k^{\text{Bkg}_\text{i}} = \mathcal{N}_k^{\text{SM}+\sum_\text{i}\text{Bkg}_\text{i}}. \tag{8}$$

The error $\delta\mathcal{N}_k^{\text{SM}}$ can be defined as

$$\delta\mathcal{N}_k^{\text{SM}+\sum\text{Bkg}_\text{i}} = \sqrt{\mathcal{N}_k^{\text{SM}+\sum_\text{i}\text{Bkg}_\text{i}}\left(1 + \delta_{\text{sys}}^2\, \mathcal{N}_k^{\text{SM}+\sum_\text{i}\text{Bkg}_\text{i}}\right)}. \tag{9}$$

The χ^2 analysis due to un-correlated systematic uncertainties is studied for three representative values of δ_{sys} at 1%, 5% and 10 %, respectively. And the sensitivity of $|V_{tb}|\,\Delta f_1^L$ at 95% C.L. is found to be of the order of $\sim 10^{-3} - 10^{-2}$ with the corresponding variation of 1% - 10% in the systematic error (which includes the luminosity error). The order of the sensitivity for other anomalous couplings varies between $\sim 10^{-2} - 10^{-1}$ at 95 % C.L.

2.2.3. Errors and correlations: Further, defining the combined $\chi^2_{comb.}(f_i, f_j)$ and taking into account of luminosity error $L \equiv \beta\bar{L}, \beta = 1 \pm \Delta\beta$:

$$\chi^2_{\text{comb.}}(f_i, f_j) \to \chi^2_{\text{comb.}}(f_i, f_j, \beta) \equiv \sum_{k=1}^{m}\sum_{i=0}^{n}\sum_{j=0}^{n}(f_i - \bar{f}_i)\left[V^{-1}\right]_{ij}^k (f_j - \bar{f}_j) + \left(\frac{\beta_k - 1}{\Delta\beta_k}\right)^2, \tag{10}$$

the sensitivity of $|V_{tb}|\,\Delta f_1^L \sim 10^{-2}$ and for other couplings it is $\sim 10^{-4}$ for $\Delta\beta \geq 5\%$.

Our analysis shows that the anomalous Wtb vertex at the LHeC can be probed to a very high accuracy in comparison to all existing limits.

3. The Higgs boson at the LHeC

As mentioned in the introduction, the Higgs boson searches are of utmost importance for all past and present colliders. However, some properties need to be measured accurately and in this respect future colliders also play a very important role. In Ref. [2] the authors studied the $hb\bar{b}$ coupling at the LHeC, and they demonstrated that the requirement of forward jet tagging in charged current events strongly enhances the signal-to-background ratio. The charged current process at the LHeC is W-vector boson fusion and hence one could measure hWW coupling strength as well. CP properties of the Higgs boson can be determined by considering an effective five-dimensional vertex, given as [3],

$$\Gamma_{\mu\nu}(p, q) = \frac{g}{M_W}\left[\lambda\left(p \cdot q g_{\mu\nu} - p_\nu q_\mu\right) + i\lambda' \epsilon_{\mu\nu\rho\sigma} p^\rho q^\sigma\right], \tag{11}$$

where λ and λ' are the effective coupling strengths for the anomalous CP-conserving and the CP-violating operators, respectively. They have shown that the azimuthal angle between missing energy and non-b jet $\Delta\phi_{MET-J}$ is a powerful and unambiguous probe of anomalous hWW couplings, both for CP-conserving and violating type.

Process	CC (fb)	NC (fb)	PHOTO (fb)
Signal:	2.40×10^{-1}	3.95×10^{-2}	3.30×10^{-6}
$b\bar{b}b\bar{b}j$:	8.20×10^{-1}	$3.60 \times 10^{+3}$	$2.85 \times 10^{+3}$
$b\bar{b}jjj$:	$6.50 \times 10^{+3}$	$2.50 \times 10^{+4}$	$1.94 \times 10^{+6}$
$zzj(z \to b\bar{b})$:	7.40×10^{-1}	1.65×10^{-2}	1.73×10^{-2}
$t\bar{t}j$(hadronic):	3.30×10^{-1}	$1.40 \times 10^{+2}$	$3.27 \times 10^{+2}$
$t\bar{t}j$(semi-leptonic):	1.22×10^{-1}	$4.90 \times 10^{+1}$	$1.05 \times 10^{+2}$

Table 3: Cross sections (in fb): $E_e = 60$ GeV, $E_p = 50$ TeV, $j = gu\bar{u}d\bar{d}s\bar{s}c\bar{c}$. Initial cuts: $|\eta| \leq 10$ for jets, leptons and b, $P_T \geq 10$ GeV, $\Delta R_{\rm min} = 0.4$ for all particles.

4. Double Higgs boson production at the FCC-he

Further plans for a high-energy LHC provides 50 TeV protons and hence the LHeC center of mass energy could be increased up to ~ 3.5 TeV with 60 GeV electrons, named as the FCC-he. This energy provides opportunities to probe the Higgs self coupling strength $g^{(1)}_{hhh}$ through double Higgs boson production. In this work we consider the speculated detector parameters and cut-based analysis to get charged-current signals with respect to all possible charged/neutral-current and photo-production backgrounds [4]. A statistical analysis is also performed to find the sensitivity of $g^{(1)}_{hhh}$ with other effective couplings described with the following effective Lagrangian:

$$\mathcal{L}^{(3)}_{hhh} = \frac{m_h^2}{2v}(1 - g^{(1)}_{hhh})h^3 + \frac{1}{2} g^{(2)}_{hhh} h\partial_\mu h \partial^\mu h, \tag{12}$$

$$\mathcal{L}^{(3)}_{hWW} = -\frac{g}{2m_W} g^{(1)}_{hWW} W^{\mu\nu} W^\dagger_{\mu\nu} h - \frac{g}{m_W}\left[g^{(2)}_{hWW} W^\nu \partial^\mu W^\dagger_{\mu\nu} h + \text{h.c.}\right] - \frac{g}{2m_W} \tilde{g}_{hWW} W^{\mu\nu} \tilde{W}^\dagger_{\mu\nu} h, \tag{13}$$

$$\mathcal{L}^{(4)}_{hhWW} = -\frac{g^2}{4m_W^2} g^{(1)}_{hhWW} W^{\mu\nu} W^\dagger_{\mu\nu} h^2 - \frac{g^2}{2m_W^2}\left[g^{(2)}_{hhWW} W^\nu \partial^\mu W^\dagger_{\mu\nu} h^2 + \text{h.c.}\right] - \frac{g^2}{4m_W^2} \tilde{g}_{hhWW} W^{\mu\nu} \tilde{W}^\dagger_{\mu\nu} h^2. \tag{14}$$

Here $g^{(1)}_{hhh}$ is defined such that it appears as a multiplicative constant to λ_{SM} i.e; $\lambda \to g^{(1)}_{hhh} \lambda_{SM}$ in the potential for electroweak symmetry breaking in the SM:

$$V(\Phi) = \mu^2 \Phi^\dagger \Phi + \lambda(\Phi^\dagger \Phi)^2 \to \frac{1}{2} m_h^2 h^2 + \lambda v h^3 + \frac{\lambda}{4} h^4, \tag{15}$$

with $\lambda = \lambda_{\rm SM} = m_h^2/(2v^2) \approx 0.13$. The effective couplings $g^{(1)}_{hhh}$, $g^{(2)}_{hhh}$, $g^{(1)}_{hWW}$, $g^{(2)}_{hWW}$, $g^{(1)}_{hhWW}$, and $g^{(2)}_{hhWW}$ are CP-conserving, whereas $\tilde{g}_{hWW}$, $\tilde{g}_{hhWW}$ are CP-violating effective couplings. m_h and m_W are respectively masses of the Higgs and W-bosons, $W^{\mu\nu} = \partial^\mu W^\nu - \partial^\nu W^\mu$.

4.1. Cross section, Detector setup and cut-based analysis

Fiducial cross sections for signal and backgrounds, before cut-based analysis, are shown in Table 3. For signal we consider the charged current process $pe^- \to \nu_e hhj, h \to b\bar{b}$. Photo-production backgrounds are very important for signals, other charged/neutral-current backgrounds, and those backgrounds estimated through "Equivalent photon approximation structure functions"; which is calculated with the "Improved Weizsaecker-Williams formula" [5].

Journal of Physics: Conference Series **645** (2015) 012005 doi:10.1088/1742-6596/645/1/012005

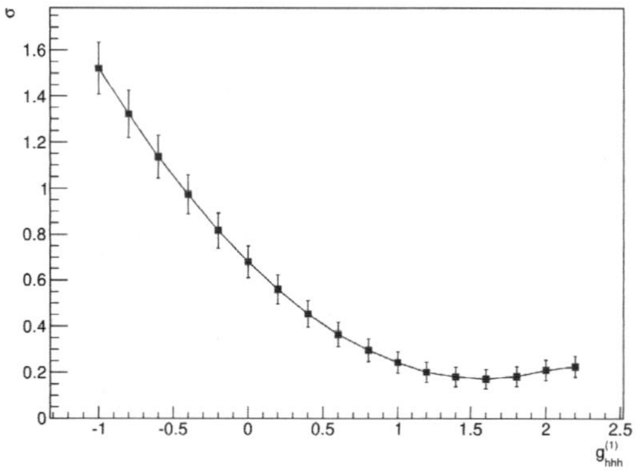

(a)

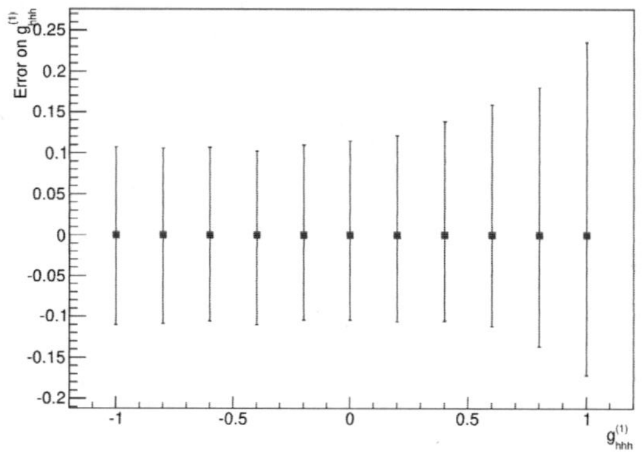

(b)

Figure 1: (a) Variation of cross section of the signal process $\sigma(pe^- \to \nu_e hhj, h \to b\bar{b})$ with respect to $g^{(1)}_{hhh}$ with error bar at each value of $g^{(1)}_{hhh}$, (b) local error through linear interpolation at each value of $g^{(1)}_{hhh}$.

In the detector setup, the maximum rapidity $|\eta|$ range is up to 7. For the b-tagging, the jets with $|\eta| < 5$ and transverse momentum $p_T > 15$ GeV is taken. The fake rate for a c-initiated jet and a light jet to the b-jet is 10% and 1% respectively. The weight corresponding to the b-tagging efficiency or fake rate is assigned to each event. Furthermore, the following cut flows are taken for analysis:

- Select 4 b + 1-jet: $p_T^{jet} > 20$ GeV, $|\eta| < 7$ for *non-b*-jets, $|\eta| < 5$ for b-jets. The four b jets must be well separated within $\Delta R > 0.7$ [1] in case of overlapped truth matching in the b-tagging.
- Rejecting leptons with $p_T^{e^-} > 10$ GeV (to suppress the neutral-current process).
- $\eta_{forward-jet} > 4.0$, the forward jet as defined as the *non-b*-jet which has the largest p_T after selecting at least 4 b-jets.
- $E_T^{miss} > 40$ GeV and $\Delta\Phi_{E_T^{miss},leadingjet} > 0.4$, $\Delta\Phi_{E_T^{miss},subleadingjet} > 0.4$ [2].
- Pair the four b-jets into two pairs and calculate the invariant masses of each pair. The composition of the pair which have the smallest variance of mass to $(m_H - 40)$ GeV is chosen. The first pair is defined as $90 < M_1 < 125$ GeV, which must have the leading b-jet. The other pair is defined as $75 < M_2 < 125$ GeV.
- Choosing the invariant mass of all four b-jets greater than 280 GeV.

And the significance is calculated with a Poisson distribution [3], considering the expected signal (S) and background (B) yields at 10 ab^{-1} luminosity. After performing these cut based analyses the signal events are ~ 63 with respect to total background events ~ 35 and $s = 8.7$. A 5% systematic error is introduced into signal and backgrounds.

[1] The ΔR is defined as the distance between two objects i and j in the rapidity-azimuthal plane: $\Delta R = \sqrt{(\phi_i - \phi_j)^2 + (\eta_i - \eta_j)^2}$, where ϕ_i and η_i are the azimuthal angle and the rapidity of the object i.

[2] $\Delta\Phi_{ij}$ is the azimuthal angle difference of two objects i and j. The (sub)leading jet is the *non-b* jet defined with (second-)highest p_T.

[3] $s = \sqrt{2\left((S+B)\, log\,(1 + S/B) - S\right)}$

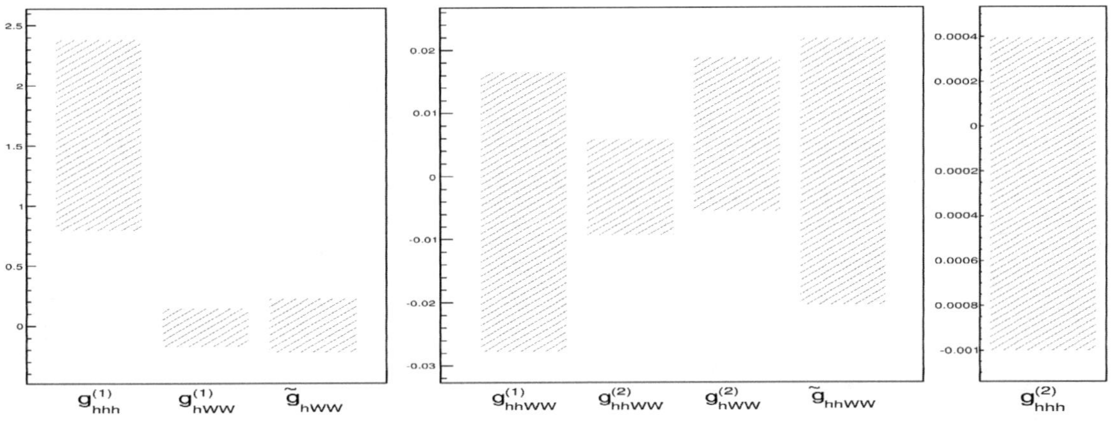

(a)

Figure 2: The limits on the coupling strength, derived at 0.4 ab^{-1}. The $g^{(2)}_{hhh}$ has only the upper limit because the cross section dependence is monotonic in this region.

4.2. Statistical analysis

Following the method given in Ref. [6], exclusion limits for $g^{(1)}_{hhh}$ are calculated. Fig. 1 shows significant behaviour of cross section variation with respect to $g^{(1)}_{hhh}$, which is expected, due to interference between resonant and non-resonant Higgs mediation in the charge-current signal process. The 95% upper limits of all effective couplings appear in the Lagrangian Eqs. (12), (13) and (14) due to cross section influence, are also calculated and shown in Fig. 2. The sensitivity of CP-even and odd HWW effective couplings, $g^{(1)}_{hWW}$ and $\tilde{g}_{hWW}$, are of the same order as $g^{(1)}_{hhh}$ $\sim 10^0$. However, for $g^{(2)}_{hhh} \sim 10^{-3} - 10^{-4}$ and $g^{(2)}_{hWW}$, $g^{(1,2)}_{hhWW}$, $\tilde{g}_{hhWW}$, these are of the order $\sim 10^{-2}$.

5. Conclusion

We briefly reviewed the physics potential of future e^-p colliders, speculated to build on top of the LHC, through the top quark and Higgs boson physics. And we infer that the LHeC and the FCC-he is a viable option to study top and Higgs physics, and for the measurement of related couplings to high accuracy.

Acknowledgements

MK would like to acknowledge all his collaborators, namely Bruce Mellado, Sukanta Dutta, Ashok Goyal, Alan Cornell, Xifeng Ruan and Rashidul Islam, as well as the organisers of the HEPP workshop 2015.

References

[1] S. Dutta, A. Goyal, M. Kumar and B. Mellado, arXiv:1307.1688 [hep-ph].
[2] T. Han and B. Mellado, Phys. Rev. D **82**, 016009 (2010) [arXiv:0909.2460 [hep-ph]].
[3] S. S. Biswal, R. M. Godbole, B. Mellado and S. Raychaudhuri, Phys. Rev. Lett. **109**, 261801 (2012) [arXiv:1203.6285 [hep-ph]].
[4] In preparation: M. Kumar, X. Ruan, A. Cornell, R. Islam, M. Klein, U. Klein, B. Mellado
[5] V. M. Budnev, I. F. Ginzburg, G. V. Meledin and V. G. Serbo, Phys. Rept. **15** (1975) 181.
[6] G. Cowan, K. Cranmer, E. Gross and O. Vitells, Eur. Phys. J. C **71** (2011) 1554 [Erratum-ibid. C **73** (2013) 2501] [arXiv:1007.1727 [physics.data-an]].

Journal of Physics: Conference Series **645** (2015) 012006 doi:10.1088/1742-6596/645/1/012006

Finite time calculations for hard parton production relevant to the quark-gluon plasma

Ben Meiring[1] **and W A Horowitz**[2]

Department of Physics, University of Cape Town, Private Bag X3, Rondebosch 7701, South Africa

E-mail: [1]`mrnben002@myuct.ac.za` and [2]`wa.horowitz@uct.ac.za`

Abstract. AdS/CFT computations have been used to describe the energy loss of QCD-like particles moving through a strongly coupled plasma, but little is understood regarding the initial conditions of these jets. We use the Schwinger-Keldysh finite-time formalism applied to an interacting scalar field theory to derive perturbative expressions detailing the system which exists during the initial stages of a high energy collision.

In this paper we calculate $\langle\phi\rangle(\mathrm{x})$ for a scalar Yukawa model, demonstrate the finiteness of the energy momentum tensor for $\lambda\phi^4$ to leading order, and derive an expression for the conditional expectation value of operators to aid in the description jet-like behaviour in interacting theories.

1. Introduction

The intention of this research is to understand the behaviour of particles in an interacting theory as a function of spacetime, during the initial moments after a high energy collision. Asymptotic freedom in QCD suggests that a perturbative approach with weak coupling will be well justified in this regime. A natural object to study is the expectation value of the energy momentum tensor. This can be done in the Schwinger-Keldysh formalism, matching the full interacting theory states to asymptotically free states at $t = \infty$ in the interaction picture. We will refer to these asymptotic states as $|\mathrm{in}\rangle$ states. For some operator $\hat{O}(t, \vec{x})$ the expectation value is given by

$$\langle \hat{O}_{\text{Heisenberg}} \rangle (t) = \left\langle \underset{\rightarrow}{T} \exp\left(i \int_{-\infty}^{t} dz_1\, \hat{H}_{\mathrm{I}}^{-}(z_1) \right) \hat{O}_{\mathrm{I}}(t) \underset{\leftarrow}{T} \exp\left(-i \int_{-\infty}^{t} dz_1\, \hat{H}_{\mathrm{I}}^{+}(z_1) \right) \right\rangle \quad (1)$$

The operator here is spacetime dependent, and so information regarding the evolution of the system can be retrieved. The $+$ and $-$ superscripts are used to disintguish the Hamiltonians of the time ordered and anti-time ordered exponentials, but the same Hamiltonians are used. It can be shown that this object is equivalent to the contour ordered exponential

$$\left\langle T_C \left(e^{-i \int_{-\infty}^{\infty} dz_1 \left(\hat{H}_I^{+}(z_1) - \hat{H}_I^{-}(z_1) \right)} \hat{O}_I(t) \right) \right\rangle, \quad (2)$$

where T_C is the contour ordering operator which orders the fields by their position along the Schwinger-Keldysh contour given below.[1]

[1] The superscript $+$ indicates a field on the top path of the contour, which will be evaluated before any field with index $-$ which is located on the bottom path.

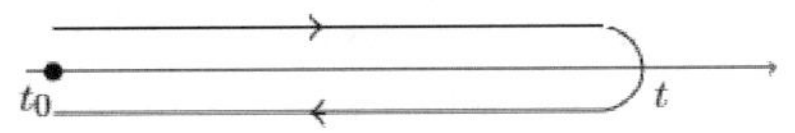

Figure 1: Time evolution indicated by the Schwinger-Keldysh contour.

This system can be solved in a similar way to usual diagrammatic calculations, now with four possible propagators due to contractions between $+$ and $-$ field operators. For some field $\phi(x)$ the contour ordered contractions $\langle 0|T_C\{\phi^i(x_1)\phi^j(x_2)\}|0\rangle$ will give propagator contributions of the form

$$D^{ij}(x_1,x_2)=\begin{cases}\langle 0|\underleftarrow{T}\{\phi(x_1)\phi(x_2)\}|0\rangle & \text{if } i,j=+,+\\ \langle 0|\{\phi(x_2)\phi(x_1)\}|0\rangle & \text{if } i,j=+,-\\ \langle 0|\{\phi(x_1)\phi(x_2)\}|0\rangle & \text{if } i,j=-,+\\ \langle 0|\underrightarrow{T}\{\phi(x_1)\phi(x_2)\}|0\rangle & \text{if } i,j=-,-\end{cases}$$

Important relations $D_R(x_1-x_2)=\sum_{m=\{+,-\}} mD^{+m}(x_1,x_2)$ and $\sum_{n,m=\{+,-\}} nmD^{nm}(x_1,x_2)=0$ can be derived through this definition.

2. Expectation of $\phi(x)$

To gain an intuition for spacetime dependent Schwinger-Keldysh calculations, we find $\langle\phi(x)\rangle$. For now we will drop the T_C path ordering operator and assume that every expectation value is contour ordered. First we note that $\langle \text{in}_1|\text{in}_2\rangle \sim \delta_{\text{in}_1,\text{in}_2}$ in Schwinger-Keldysh, which can be seen by finding the expectation of the identity operator. In general we can choose the operator $\hat{O}$ to be evaluated on the $+$ or $-$ contour without changing the result. Choosing $\phi(x)$ on the $+$ contour, we will have

$$\langle\phi(x)\rangle_{\text{Heisenberg}}=\sum_{n=0}^{\infty}\frac{(-i)^n}{n!}\langle\phi^+(x)\prod_{i=1}^{n}\sum_{m_i=\{+,-\}} m_iH^{m_i}\rangle_{\text{interaction}} \tag{3}$$

If we choose the incoming states to be the same $|\text{in}_1\rangle$ then contracting $\phi(x)$ with an external state will allow us to factorize the operator out. Thus the rest of diagram will depend on the expectation $\langle \text{in}_1|\mathbb{I}|\text{in}_2\rangle$ with $|\text{in}_1\rangle\neq|\text{in}_2\rangle$. Thus these diagrams will contribute 0, and $\phi(x)$ will only contract with $\phi(z_i)$ operators from the interaction Hamiltonians.

Let $H^{m_i}=\int d^4z\,\left(g\psi\phi\psi^{m_i}-\delta_g\psi\phi\psi^{m_i}+\frac{1}{2}\psi\left(\delta_\psi\partial^2-\delta_{m_\psi}\right)\psi^{m_i}+\frac{1}{2}\phi\left(\delta_\phi\partial^2-\delta_{m_\phi}\right)\phi^{m_i}\right)$. The counter terms here simply cancel the divergences that come from the loop integrals, so we will not consider their effect in too much detail. Focusing on the $g\psi\phi\psi^{m_i}$ term, we explicitly write out the possible contractions.

$$\langle\phi(x)\rangle_{\text{Heisenberg}}=\sum_{n=0}^{\infty}\frac{(-ig)^n}{n!}\langle\phi^+(x)\int d^4z_1...d^4z_n\sum_{m_i=\{+,-\}} m_1\psi\phi\psi^{m_1}...\sum_{m_n=\{+,-\}} m_n\psi\phi\psi^{m_n}\rangle$$

$$=\sum_{n=0}^{\infty}\frac{(-ig)^n n}{n!}\int d^4z_n\langle\sum_{m_n=\{+,-\}} m_n\overline{\phi^+(x)\phi^{m_n}}(z_n)\left(\psi\psi^{m_n}(z_n)\right)\prod_{i}^{n-1}\int d^4z_i\sum_{m_i=\{+,-\}} m_i\psi\phi\psi^{m_i}\rangle$$

$$=(-ig)\int d^4z\left(\sum_{m=\{+,-\}} mD^{+m}(x,z)\right)\sum_{n=0}^{\infty}\frac{(-ig)^{n-1}}{(n-1)!}\langle\left(\psi\psi^m(z)\right)\prod_{i}^{n-1}\int d^4z_i\sum_{m_i=\{+,-\}} m_i\psi\phi\psi^{m_i}\rangle$$

$$=(-ig)\int d^4z\left(\sum_{m=\{+,-\}} mD^{+m}(x,z)\right)\langle\psi\psi^m(z)\rangle_{\text{Heisenberg}}$$

Journal of Physics: Conference Series **645** (2015) 012006 doi:10.1088/1742-6596/645/1/012006

Including the counter terms in this analysis gives us the same result, now with all loop contributions being renormalized. $\langle \psi\psi^m(z)\rangle_{\text{Heisenberg}}$ is just the expectation value of some operator. It does not matter if this operator is evaluated on the $+$ or $-$ contour, the answer will be the same through the m sum.

$$\begin{aligned}\langle \phi(x)\rangle_{\text{Heisenberg}} &= (-ig)\int d^4z \left(\sum_{m=\{+,-\}} mD^{+m}(x,z)\right)\langle \psi\psi(z)\rangle_{\text{Heisenberg}} \\ &= (-ig)\int d^4z\, D_R(x-z)\langle \text{in}|\psi\psi(z)|\text{in}\rangle_{\text{Heisenberg}}\end{aligned}$$

2.1. Leading order emission from a single particle

The above result will hold in arbitrary dimensions.

$$\langle \phi(x)\rangle = -ig\int d^{n+1}z\, D_R(x-z)\langle \psi\psi(z)\rangle \tag{4}$$

Notice that this is the solution to the equations of motion given by classically coupled lagrangian $\mathcal{L} = \frac{1}{2}\partial_\mu\phi\partial^\mu\phi - \frac{1}{2}m_\phi^2\phi^2 - g\phi\langle\psi\psi\rangle$. In general if we are able to determine the expectation $\langle\psi\psi\rangle$, we will have this expectation exactly. This gives this quantity the interpretation of being a propagation of the ϕ field from the current given by the expectation of the square of the ψ. In weak coupling this case be considered as a statistical ensamble of the possible virtual/real particles off which the ϕ field can be emitted.

Suppose we have $|\text{in}\rangle = |\psi\rangle$. To leading order we will find $\langle\psi|\psi\psi|\psi\rangle(z) = 2\langle\overbrace{\psi|\psi}\overbrace{\psi|\psi}\rangle(z)$ Using wavepackets $\tilde{\psi}(k)$ we can represent the assumptotic states as

$$|\psi\rangle = \int \frac{d^nk}{(2\pi)^{n/2}\sqrt{2E_k}}\tilde{\psi}(k)|\vec{k}\rangle. \tag{5}$$

with normalisation condition $\int \frac{d^nk}{(2\pi)^n}|\tilde{\psi}(k)|^2 = 1$. We find

$$\langle\psi|\psi\psi|\psi\rangle(z) = \int \frac{d^nk_1d^nk_2}{(2\pi)^n\sqrt{E_{k_1}E_{k_2}}}\tilde{\psi}(k_1)\tilde{\psi}^*(k_2)e^{-iz(k_1-k_2)}. \tag{6}$$

Then

$$\langle\phi\rangle(x) = g\int \frac{d^nk_1d^nk_2}{(2\pi)^n\sqrt{E_{k_1}E_{k_2}}}\tilde{\psi}(k_1)\tilde{\psi}^*(k_2)\frac{e^{-ix(k_1-k_2)}}{(k_1-k_2)^2 - m_\phi^2 + i\epsilon(k_1^0-k_2^0)} \tag{7}$$

For an analytic result we take $\tilde{\psi}(k) = \frac{(2\pi)^{n/2}}{(2\pi\alpha^2)^{n/4}}e^{-\frac{1}{4}(\vec{k})^2/\alpha^2}$ Effectively there is only a significant contribution to the integral when $\vec{k}_1^2 + \vec{k}_2^2 << \alpha^2$. As $\alpha \to 0$ the wavepackets become sharply peaked and this becomes a smaller region. If $m_\psi >> \alpha^2$ then $E_q \approx m_\psi$ through the integral.[2] Thus

$$\langle\psi|\psi\psi|\psi\rangle(z) \approx \frac{1}{m_\psi}\left|\int \frac{d^nk}{(2\pi)^{n/2}}\tilde{\psi}(k)e^{-i\vec{z}\cdot\vec{k}}\right|^2 = \frac{1}{m_\psi}|\tilde{\psi}(\vec{z})|^2 \tag{8}$$

Note that where the time is small in comparison to m_ψ, the time dependence drops out of the expression. We are left with a static, stable source over this time interval. Thus

$$\langle\phi(x)\rangle \approx -i\frac{g}{m_\psi}\int d^{n+1}z\, D_R(x-z)|\tilde{\psi}(\vec{z})|^2. \tag{9}$$

[2] This argument will hold true for any peaked wavepacket with width parameterized by α.

Journal of Physics: Conference Series **645** (2015) 012006 doi:10.1088/1742-6596/645/1/012006

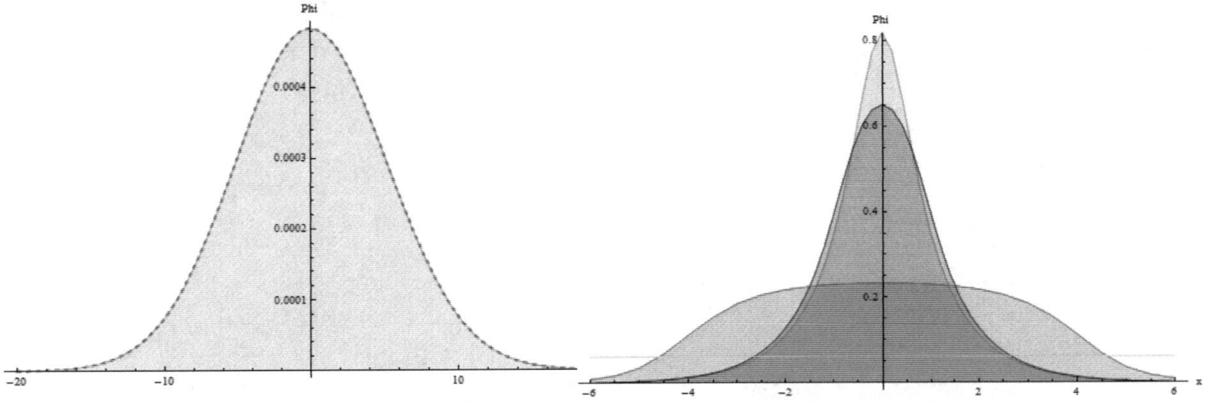

(a) Comparison between numerical and analytic results for $\langle\phi(x)\rangle$ in $1+1$ dimensions, for large m_ψ

(b) Time evolution of $\langle\phi(x)\rangle$ in $1+1$ dimensions

Figure 2: Numerical Solutions in 1+1 D for $\langle\phi(x)\rangle$.

For our chosen wavepackets, $|\tilde{\psi}(\vec{z})|^2 = 2^n\sqrt{2\pi\alpha^2}^n e^{-2\alpha^2\vec{z}^2}$. Making the substitution $2\alpha^2 = \frac{1}{4\tau}$, choosing $n = 3$ and integrating the expression over z^0, we will find

$$\langle\phi(x)\rangle \approx (2\pi)^3 \frac{g}{m_\psi} \int d^3z \, \frac{e^{-m_\phi|\vec{z}|}}{4\pi|\vec{z}|} \frac{e^{-\frac{(\vec{z}-\vec{x})^2}{4\tau}}}{\sqrt{4\pi\tau}^3} \tag{10}$$

We can think of this as the solution of the heat equation with initial condition given by $\frac{e^{-m_\phi|\vec{x}|}}{4\pi|\vec{x}|}$, the Yukawa potential. τ is some function of the width of the asymptotic wavepackets, as we localize this momenta (or spread out the concentration of the source in position space) we smear out the concentration of $\langle\phi(x)\rangle$ given at the origin. For non-zero τ this contribution will be finite. Similar arguments will hold in arbitrary dimensions. Far from the origin, the "diffusion" has not yet had a noticable effect, and the result will approach what we would expect in the limit $\tau = 0$. In $n = 1$ the result can be written in closed form.

$$\langle\phi(x)\rangle \approx \frac{g}{m_\phi m_\psi}\frac{\pi}{2}e^{\frac{m_\phi^2}{8\alpha^2}}\left(e^{-m_\phi|\vec{x}|}\mathrm{Erfc}\left(\frac{m_\phi - 4|\vec{x}|\alpha^2}{2\sqrt{2\alpha^2}}\right) + e^{m_\phi|\vec{x}|}\mathrm{Erfc}\left(\frac{m_\phi + 4|\vec{x}|\alpha^2}{2\sqrt{2\alpha^2}}\right)\right) \tag{11}$$

3. Brief look at the improved energy momentum tensor

In general, expecations of operators in QFT are not finite, and in particular $\langle T_{\mu\nu}\rangle = \langle\sum_i \partial_\mu\phi_i\partial_\nu\phi_i - g_{\mu\nu}\mathcal{L}\rangle$ diverges order by order even after standard renormalization techniques. One must instead compute the improved energy momentum tensor[4] $\langle\Theta_{\mu\nu}\rangle = \langle T_{\mu\nu}\rangle + \langle\frac{1}{4}\frac{n-2}{n-1}\left(\partial_\mu\partial_\nu - g_{\mu\nu}\partial^2\right)\sum_i \phi_i^2\rangle$ for a quantity that is finite to at least one loop order.[3] Notice that this quantity is still satisfies $\partial^\mu\Theta_{\mu\nu} = 0_\nu$ and will produce the same result as $T_{\mu\nu}$ when integrated over all space.

Consider scalar $\lambda\phi^4$ theory with $|\mathrm{in}\rangle = |\phi\rangle$. The divergent contribution (when $\Theta_{\mu\nu}$ is evaluated on the + contour) is given exclusively by the time ordered propagators, so we will ignore all other contributions when showing finiteness. Define $M_{\mu\nu}$ as the sum of all time ordered momentum space diagrams contributing to $\langle\Theta_{\mu\nu}\rangle$ at order λ. Then we can find

$$M_{\mu\nu} = 24i\lambda \int \frac{d^dq}{(2\pi)^4} \frac{(q-k)_\mu q_\nu - \frac{1}{2}g_{\mu\nu}((q-k)q - m^2) + \frac{1}{4}\frac{n-2}{n-1}(k_\mu k_\nu - g_{\mu\nu}k^2)}{((q-k)^2 - m^2)(q^2 - m^2)} \tag{12}$$

[3] n here is the number of spatial dimensions.

With $k = k_1 - k_2$ being the different in momenta of the bra and ket asymptotic states, and m the mass of the ϕ field. Working through this expression we can find the divergent part to be given by

$$\frac{24\lambda}{(4\pi)^{d/2}} \int_0^1 dx \left(\left(x(1-x) - \frac{1}{6} \right) (k_\mu k_\nu - g_{\mu\nu} k^2) \right) \frac{1}{2 - \frac{d}{2}} = 0. \tag{13}$$

Thus we can make sense of the $\Theta_{\mu\nu}$ operator to leading order in λ.

4. Conditional Expectation Value

To understand $\langle \Theta_{\mu\nu} \rangle(x)$ for a jet we propose choosing an initial state of two particles. After a two-particle collision, an infinite number of final states are kinematically accessible. Thus $\langle \Theta_{\mu\nu} \rangle$ will be some average value of these possible states. To restrict this expectation value with additional information we define the conditional expectation value. This is the expectation given both the initial and final states.

4.1. Derivation

Let r_n be an eigenvalue of the operator $\hat{R}$ and Δ a set of some of these eigenvalues. The probability that a measurement of $\hat{R}$ on some state will yield r_n is 1, what is the probability that it will yield $r_n \in \Delta$? We define

$$M_R(\Delta) = \sum_{r_n \in \Delta} |r_n\rangle\langle r_n| \tag{14}$$

This is the projection operator onto states given by Δ. Define density matrix $\rho = \sum_b \rho_b |b\rangle\langle b|$ with property $\sum_b \rho_b = 1$. The probability that a measurement will yield a result in Δ is the average value of the projection operator. This can be motivated by noting that if Δ_c is a complete set, $M_R(\Delta_c) = 1$. Then $\text{Tr}\{\rho M_R(\Delta_c)\} = 1$. If Δ is a smaller subset, the expectation of this operator now excludes the contribution from the missing states. Thus we say that

$$P(\{R \in \Delta\}|\rho) = \text{Tr}\{\rho M_R(\Delta)\} \tag{15}$$

We now want to consider two observables Q and R represented respectively by operators $\hat{Q}$ and $\hat{R}$ with eigenvectors given by $|q_n\rangle$ and $|r_n\rangle$ with $q_n \in \Gamma$ and $r_n \in \Delta$.

To define the conditional probability, to need to know the probability that a state corresponding to $q_n \in \Gamma$ will be measured, given that a state corresponding to $r_n \in \Delta$ has been measured. This is simply

$$P(\{Q \in \Gamma\}|\tilde{\rho}) = \text{Tr}\{\tilde{\rho} M_Q(\Gamma)\} \tag{16}$$

where $\tilde{\rho}$ is the state of the system after a measurement of $\hat{R}$. How do we find $\tilde{\rho}$? If a state $\tilde{\rho}$ is such that a measurement will yield $r_n \in \Delta$ with certainty, then $\tilde{\rho} \sim M_R(\Delta)\hat{A}M_R(\Delta)$. Enforcing that $\text{Tr}\{\tilde{\rho}\} = 1$ provides a normalization of $\text{Tr}\{M_R(\Delta)\hat{A}M_R(\Delta)\}$.

Suppose we collapsed the state into $\tilde{\rho}(0)$ immediately after preparing the state in $\rho(0)$. Then we would find $\hat{A} = \rho(0)$. Define $\hat{\rho}(t)$ as the time evolved $\hat{\rho}(0)$ state. Then

$$\tilde{\rho}(t) = \frac{M_R(\Delta)\hat{\rho}(t)M_R(\Delta)}{\text{Tr}\{M_R(\Delta)\hat{\rho}(t)M_R(\Delta)\}} = \frac{M_R(\Delta)\hat{\rho}(t)M_R(\Delta)}{\text{Tr}\{\hat{\rho}(t)M_R(\Delta)\}} \tag{17}$$

where we have used the cyclicity of the trace and the fact that $M_R(\Delta)$ is a projection operator. Now we can write

$$P(\{Q \in \Gamma\}|\tilde{\rho}) = \frac{\text{Tr}\{M_R(\Delta)\hat{\rho}(t)M_R(\Delta)M_Q(\Gamma)\}}{\text{Tr}\{\hat{\rho}(t)M_R(\Delta)\}} = \frac{\text{Tr}\{\hat{\rho}(t)M_R(\Delta)M_Q(\Gamma)M_R(\Delta)\}}{\text{Tr}\{\hat{\rho}(t)M_R(\Delta)\}}. \tag{18}$$

This is the conditional probability that a state corresponding to $q_n \in \Gamma$ will be measured, given that a state corresponding to $r_n \in \Delta$ has been measured.

The conditional expecation value of an operator $\hat{\Theta}$ is defined

$$E(\hat{\Theta}, \{R \in \Delta\}|\rho) = \sum_{q_n \in \Gamma} \Theta_{q_n} P(Q \in \{q_n\}|\tilde{\rho}) \tag{19}$$

the sum of the eigenvalues Θ_{q_n} of $\hat{\Theta}$ weighted by the conditional probability. For $\Theta = 1$ this reduces to $E(1, \{R \in \Delta\}|\rho) = P(Q \in \Gamma|\tilde{\rho})$ as expected. We will take Γ to be a complete set so that any eigenvalue of $\hat{\Theta}$ is accessible.

We can write $\hat{\Theta} = \sum_{q_n \in \Gamma} \Theta_{q_n} |q_n\rangle\langle q_n|$, allowing us to express the conditional probability as

$$E(\hat{\Theta}, \{R \in \Delta\}|\rho) = \sum_{q_n \in \Gamma} \Theta_{q_n} \frac{\mathrm{Tr}\{\hat{\rho}(s) M_R(\Delta)|q_n\rangle\langle q_n| M_R(\Delta)\}}{\mathrm{Tr}\{\hat{\rho}(s) M_R(\Delta)\}} = \frac{\mathrm{Tr}\{\hat{\rho}(s) M_R(\Delta)\hat{\Theta}(t) M_R(\Delta)\}}{\mathrm{Tr}\{\hat{\rho}(s) M_R(\Delta)\}} \tag{20}$$

For the situations in which we are interested, we will set $s = \infty$ (the time at which the state $M_R(\Delta)$ is measured). We can consider $\hat{\Theta}$ to be given at some arbitrary time t because we have used a general eigenbasis given by $|q_n(t)\rangle$. Writing the density matrix in terms of some wavefunction for the system we will have

$$E(\hat{\Theta}, \{R \in \Delta\}|\rho) = \frac{\langle \Psi | M_R(\Delta)\hat{\Theta}(t) M_R(\Delta)|\Psi\rangle(s)}{\langle \Psi | M_R(\Delta)|\Psi\rangle(s)} \tag{21}$$

For an interacting theory, this expectation value can again be related to the asymptotic free field eigenvectors through the interaction picture, and perturbation theory can be done in the usual way. The new contour is given in the diagram provided.

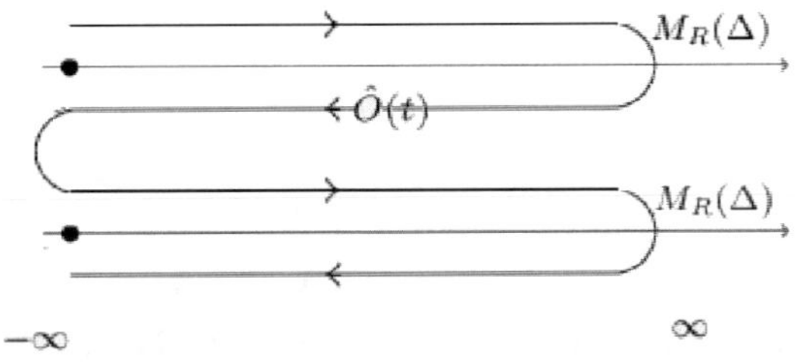

Figure 3: Time evolution indicated by a modified contour with various measurements

5. Conclusion

We have made progress towards building a spacetime description phenomena in quantum field theory, in particular towards finding the structure of jets in the perturbative regime. We have shown how calculations can be done in the case of $\langle \phi(x) \rangle$ for a simple scalar theory, and have demonstrated how to make sense of the energy momentum tensor in QFT through the improvement term. To extend this work we intend to calculate the conditional expectation value of $\Theta_{\mu\nu}$ given a simple $\psi\psi \to \psi\psi$ scalar particle collision. We also plan to derive similar results for theories containing gauge fields and fermions.

References

[1] Peskin M E and Schroeder D V 1995 *An Introduction to Quantum Field Theory* (Reading, MA: Addison-Wesley Pub. Co.)
[2] Le Bellac M 1996 *Thermal Field Theory* (Cambridge: Cambridge University Press) 35-85
[3] Collins J 1984 *Renormalization* (Cambridge: Cambridge University Press)
[4] Callan C, Coleman S and Jackiw R. 1970 *Annals of physics* **59** 42-73

Jets of light hadrons via AdS/CFT correspondence

R Morad[1] and W A Horowitz[2]

Department of Physics, University of Cape Town, Private Bag X3, Rondebosch 7701, South Africa

E-mail: [1]`razieh.morad@uct.ac.za`, [2] `wa.horowitz@uct.ac.za`

Abstract.
Within the expanding fireball formed at heavy ion collision, jets are produced which probes the QGP. Analysis the energy loss of these energetic partons as they travel throw QGP may reveal extremely valuable information about the dynamics of the plasma and exhibit distinctive properties such as jet-quenching. The "AdS/CFT correspondence" which imposes the duality between the gauge theory and gravity is a novel tool provides valuable insight into the strongly coupled plasma. The most important result of AdS/CFT is calculating the value of shear viscosity to entropy density ratio which is in remarkable agreement with the hydrodynamics predictions. We study the energy loss rate of light quarks via AdS/CFT correspondence in both static and expanding plasma. In the hope of making contact with QGP physics, we propose a novel jet prescription based on the separation of hard and soft modes in the dual theory and test the AdS/CFT approach with the latest light hadron suppression data from CMS.

1. Introduction

The spectacular measurements from the Relativistic Heavy Ion Collider (RHIC) and the Large Hadron Collider (LHC) provide compelling evidence that the matter produced in heavy ion collision is a deconfined state of QCD, Quark-Gluon Plasma (QGP) [1, 2, 3, 4], at temperatures above ~160 MeV which appears to be nearly perfect, with an extremely low viscosity-to-entropy ratio $\eta/s \sim 1/4\pi$ [5, 6]. While lattice QCD is the proper tool for understanding the static equilibrium thermodynamics of such strongly coupled plasma, it does not allow us to calculate its dynamics evolution on heavy-ion collision.

Recently, a novel tool called "the AdS/CFT correspondence" [7, 8, 9, 10, 11] provide valuable insight into the strongly coupled plasma. This conjecture state that the $\mathcal{N} = 4$ SYM theory in the large N_c and large 't Hooft coupling is dual to classical supergravity on ten-dimensional $AdS_5 \times S^5$ geometry [7]. In order to study the theory at finite temperature, one can add black hole (BH) to the geometry [8] which yields to the AdS-Sch metric. Fundamental quarks are described by open strings moving in the 10d geometry. In the large λ limit, the quantum fluctuation of string world sheet are suppressed and the dynamics of string are described by the classical string theory.

Jets are produced within the expanding fireball and probe the QGP. Analyses the energy loss of these energetic partons as they travel throw QGP may reveal extremely valuable information about the dynamics of the plasma and exhibit distinctive properties such as jet-quenching which can clearly be observed at RHIC [1, 2, 3, 4] and more recently LHC [12, 13, 14].

In this paper, we study the light quark jet energy loss in both gravitational background dual to static and expanding plasma. We propose a new prescription of jets in string theory based on the separation of hard and soft sectors. We demonstrate that the light quark jet energy loss reveal the "Bragg peak" at late times. Then we compute the nuclear modification factor of jets R^{jet}_{AA}, renormalise the quantity, and compare the results with the preliminary CMS data [15].

2. Light quark jet energy loss

According to the AdS/CFT correspondence [10], the $\mathcal{N} = 4$ SYM theory at finite temperature is dual to a $10d$ black hole geometry with the AdS-Schwarzschild (AdS-Sch) metric as follows,

$$ds^2 = \frac{L^2}{u^2}\left[-f(u)\,dt^2 + d\mathbf{x}^2 + \frac{du^2}{f(u)}\right], \tag{1}$$

where $f(u) \equiv 1 - (u/u_h)^4$ is the blackening factor and L is the AdS curvature radius. Four dimensional Minkowski coordinates are denoted by x_μ and the coordinate u is an inverse radial coordinate. So, the boundary of the AdS-Sch spacetime is at $u = 0$ and the event horizon is located at $u = u_h$. The temperature of the equilibrium SYM plasma relates to the event horizon as $T \equiv \frac{1}{(\pi u_h)}$. World sheet coordinates are as σ^a where $\tau \equiv \sigma^0$ is denoted as the timelike world sheet coordinate, while the spatial coordinate is $\sigma \equiv \sigma^1$.

Fundamental representation quarks added to the $\mathcal{N}=4$ SYM theory are dual to open strings moving in the $10d$ geometry. Addition of a $\mathcal{N}=2$ hypermultiplet to the $\mathcal{N}=4$ SYM theory is performed by adding D7 branes to the $10d$ geometry [16]. These branes extend along the radial coordinate from the boundary at $u = 0$ down to maximal coordinate at $u = u_m$ as well as fill the whole 4d Minkowski space. Also, they wrap on S^3 from the S^5 sphere. The bare mass M of quark is proportional to $1/u_m$ [17], so for massless quarks the D7 brane should fill the whole radial direction. Open strings that are attached to the D7 brane are dual to the quark-anti quark pairs on the field theory side. In the $5d$ geometry these strings can fall unimpeded toward the event horizon until their end points reach the radial coordinate u_m where the D7 brane ends. Since for sufficiently light or massless quarks $u_m > u_h$, open string end points can fall into the horizon.

We are interested in studying the back-to-back jets so we consider the configurations in which the two endpoints of string move away from each other as the total spatial momentum of the string vanishes. By choosing the appropriate frame, one half of the string has a large spatial momentum in x direction, and the other half of the string carries a large negative spatial momentum. We will limit our attention to strings which move in one direction in the R^3 space, x direction, so the embedding function of string $X^\mu(\tau,\sigma)$ will be a map to $(t(\tau,\sigma), x(\tau,\sigma), u(\tau,\sigma))$. So, the profile of an open string which is created at a point in space at time $t = t_c$ is given by

$$t(0,\sigma) = t_c\,, \quad x(0,\sigma) = 0\,, \quad u(0,\sigma) = u_c\,. \tag{2}$$

By these conditions, the string created at time t_c and by time evolution, the string evolves from a point into an extended object and the string endpoints fall toward the horizon.

The dynamics of string is governed by the Polyakov action

$$S_P = -\frac{T_0}{2}\int d^2\sigma\,\sqrt{-\eta}\,\eta^{ab}\,\partial_a X^\mu \partial_b X^\nu\, G_{\mu\nu}\,. \tag{3}$$

Variation of the Polyakov action with respect to the embedding functions X^μ lead to the equation of motion as

$$\partial_a\left[\sqrt{-\eta}\,\Pi^a_\mu\right] = \frac{1}{2}\sqrt{-\eta}\,\eta^{ab}\frac{\partial G_{\nu\rho}}{\partial X^\mu}\,\partial_a X^\nu \partial_b X^\rho, \tag{4}$$

where Π^a_μ are the canonical momentum densities associated to the string obtained from varying the action with respect to the derivatives of the embedding functions.

In order to study jets in AdS/CFT we need to define the proper objects in the dual string theory that corresponds to a jet, a slippery object even in field theory; jets are truly only defined by the algorithm used to measure them. Presumably the ideal way to compute jet observables in the dual theory is to compute the energy momentum tensor associated with a high-momentum probe and "run" a jet finding algorithm on the result. However, one can define a jet prescription in the AdS/CFT and calculate the rate of energy loss from the string profile itself.

The authors of [18] are motivated by the localization of the baryon density on the boundary which is of scale of order $\Delta x \sim 1/\pi T$ and defined jet as a part of string which is in the Δx spatial distance from the endpoint. We called this as the "$\Delta x - prescription$" of jet [18].

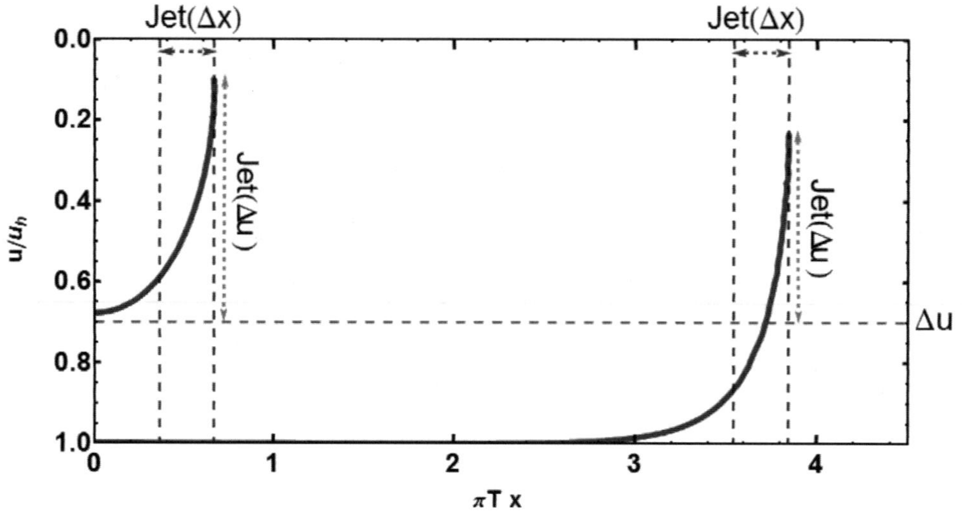

Figure 1. Illustration of the Δx and Δu prescriptions of a jet in the string theory; see text for details.

In this paper, motivated by the separation of energy scales in, e.g., thermal field theory, we propose rather a Δu prescription which we believe will ultimately provide a closer approximation to the result of a more complete calculation, figure 1. Since the radial coordinate in the string theory sets an energy scale in the field theory, in our Δu prescription the portion of the string above some cutoff $u = u^*$ in the radial direction is considered part of the jet; the portion of the string below the cutoff is considered part of the thermalized medium. By choosing any value of u above the black hole horizon as the cutoff, we regain the natural result that a jet that is thermalized no longer has detectable energy or momentum.

We evaluate the energy loss rate of jet in the radial $\sigma = u$ parametrization for both prescriptions of jet and plot in figure 2(a,b). In order to define jet using the $\Delta x - prescription$, we choose $\sigma_\kappa(t)$ as $\Delta x = 0.3/\pi T$. Our result on the energy loss rate of light quark using the $\Delta u - prescription$ shows a Bragg peak at late times which means the explosive transfer of quark energy to the plasma at late times and is consistence with the previous works [18].

Since the quark-gluon plasma produced at ultra-relativistic heavy ion collision is an expanding and cooling medium, we study the light quark jet energy loss in a time dependent gravity dual to the boost invariant flow [19]. This geometry is similar to the static black hole geometry, but the location of the horizon moves in the bulk as $\tau^{1/3}$ where τ is the proper time. Also the temperature of plasma cools as $T(\tau)\ \tau^{-1/3}$.

We calculate the energy loss rate of light quark using both $\Delta x - prescription$ and $\Delta u - prescription$ of jet. We consider the initial temperature of plasma in JP metric T_c the same

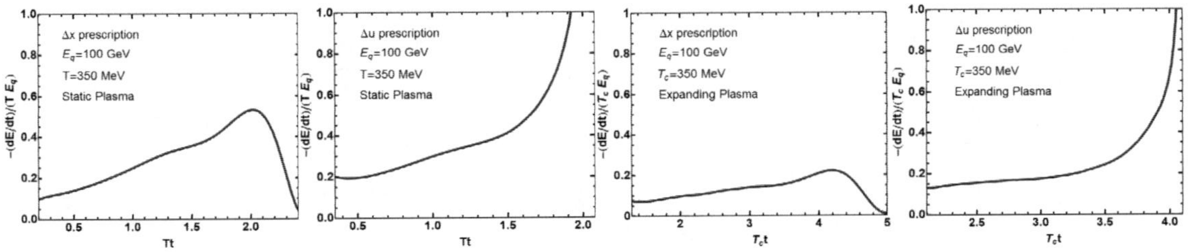

(a) (b) (c) (d)

Figure 2. The instantaneous energy loss of a light quark jet as a function of time in the AdS-Sch (a,b) and JP metrics (c,d) in the Δx prescription and Δu prescription. The normalization constant $E_q = 100$ GeV is the initial energy of the jet, which has a virtuality of 175 GeV^2, and $T = 350$ MeV is the temperature of the plasma.

as the temperature of plasma in AdS-Sch metric. The result is shown in figure 2(c,d) which demonstrates that the behavior of light quark energy loss in the JP metric is same as the AdS-Sch metric, but the distance that quark traveled before thermalizing increases.

3. Nuclear Modification Factor of Jet

We calculate an approximation of the nuclear modification factor R_{AA} for jet using our energy loss model of jet. We consider the contributions of both quark and gluon jets. We assume that the produced parton with initial energy p_T^i loses a fraction of its energy ϵ with probability $P\left(\epsilon|p_T^i, L, T\right)$ as the final energy of parton is given by $p_T^f = (1-\epsilon)p_T^i$. Also, we suppose that the the AdS energy loss is approximately independent of the initial energy [20], and the gluons loss their energy by the factor of 2 in the large N_c limit respect to the quarks.

We suppose that the production spectrum can be approximately by a power law [20], with slowly varying with respect to p_T, then we may find a simple equation for the jet nuclear modification factor as follows,

$$R_{AA}^{R\to jet}(p_T) = \left\langle \int d\epsilon\, P(\epsilon|p_T, L, T)\, \left(1-\epsilon^R\right)^{n_R(p_T)-1} \right\rangle. \tag{5}$$

For an absolutely uniform nucleus that is a 1D line, the geometric average is carried out as an integral over a line of production points with a parton that propagates through the line. In this case, $R_{AA}^{R\to jet}(p_T)$ is obtained from the below line integral [20]

$$R_{AA}^{R\to jet}(p_T) = \int_0^{L_{max}} \frac{dl}{L_{max}} \left(1-\epsilon^R(p_T, l, T)\right)^{n_R(p_T)-1}. \tag{6}$$

We calculate R_{AA} of jet by using our results of jet energy loss in both AdS-Sch and JP metric. Our results are shown in figure 3. The purple curve shows the R_{AA} obtained from the AdS-Sch metric, while the blue curve is the R_{AA} obtained from the JP metric. The AdS/CFT results of jet energy loss show an over suppression of jets in both static and expanding plasma.

The point-like initial condition falling string that we consider here is dual to creation of a pair of quark-antiquark which fly away each other in the strongly coupled plasma, interact and loss their energy. We expect that jets produced in the pp collision do not loss their energy. So, we consider the falling string with the same initial conditions in AdS_5 metric. Our results show that the string falls in the empty AdS_5. So, jet loses its energy in the vacuum! We calculate the R_{AA} for a falling string in AdS_5 metric and show in figure 3 (a) in red curve.

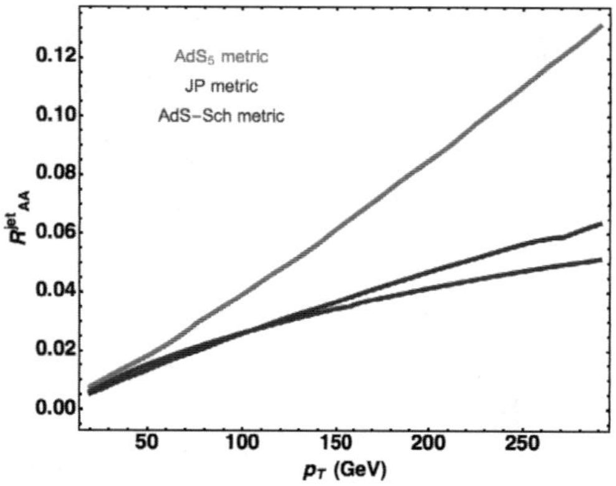

(a)

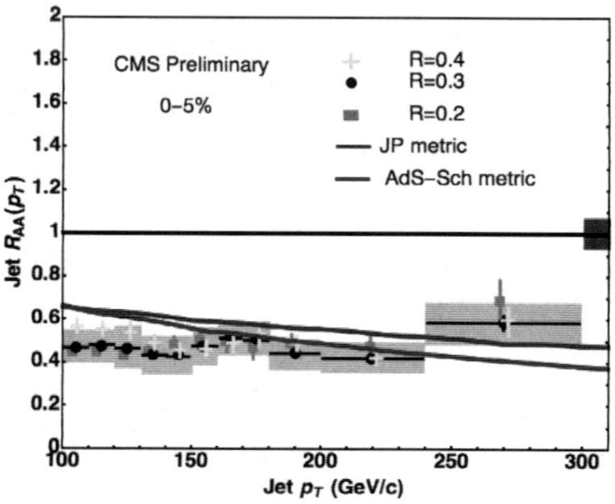

(b)

Figure 3. (a) Jet R_{AA} as a function of p_T in the most central Pb-Pb collision obtained via AdS/CFT in AdS_5 (red), JP (blue) and AdS-Sch (purple) metrics. (b) AdS/CFT jet R_{AA} as a function of p_T compared with the preliminary CMS data in different effective cone sizes for anti-k_T jets using the Bayesian unfolding method for most central Pb-Pb collision at the LHC with $\sqrt{s} = 2.76$ TeV per nucleon [15].

Since the definition of R_{AA} measured the difference between the medium and vacuum effects on jet, we define a renormalized R_{AA} in AdS/CFT as

$$R_{AA}^{renorm} = \frac{R_{AA}^{medium}}{R_{AA}^{AdS_5}} \tag{7}$$

We plot the renormalized R_{AA}^{renorm} for jets in both AdS-Sch and JP metric in figure 3 (b) and compare with the CMS preliminary data for the most central Pb-Pb collision at $\sqrt{s_{NN}} = 2.76\, GeV$ [15]. A suppression factor of 0.5 for high p_T jets is observed in central Pb-Pb collision in comparison to the pp collisions. Our results also show surprisingly agreement with CMS preliminary results on jet R_{AA}.

4. Conclusions

In this paper we have purposed a novel prescription of jet in the context of string theory and AdS/CFT correspondence. We have defined jet as a part of a falling string which lies above the radial coordinate $u^{\star}$ near the event horizon. Our motivation is defining the jet as hard partons. The instantaneous energy loss of light quark is identified with the energy flux from the point at $u(\tau, \sigma_\kappa) = u^{\star}$. We have shown that using the $\Delta u - prescription$ of jet, the light quark energy loss exhibit the Bragg peak again at late times in both static and expanding plasmas. This late time behavior of jet energy lost implies that after traveling substantial distances through the plasma, the thermalization of light quark ends with a large amount of energy transferring to the plasma which is similar to the energy loss rate of a fast charge particle moving through ordinary matter.

We considered a brick of plasma and calculated the nuclear modification factor of jet in both AdS-Sch and JP metric. We assumed that the temperature of plasma around 350 MeV at AdS-Sch metric and at initial time in JP metric. Our results show an aver suppression of jet of order of ten respect to the data. We investigated that it is because of the falling string setup at AdS space. In fact, R_{AA} of jet using the falling string in AdS_5 which is dual to jets in vacuum is not one, even though it is less than one. We introduced a renormalized R_{AA}^{renorm} by dividing the

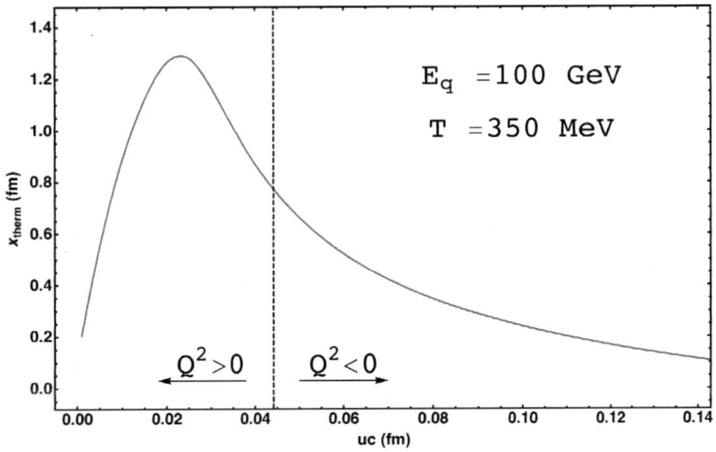

Figure 4. The maximum stopping distance of jet versus the virtuality of quark Q^2. Our result show a nontrivial dependency of the thermalization time to the initial condition of the string.

R_{AA} in the medium to the R_{AA} in the vacuum. Surprisingly, our ratio shows good agreement with the experimental data on the R_{AA}^{jet} of most central Pb-Pb collision at LHC figure 3 (b).

On the other hand, the light quark energy loss is highly depends on the initial conditions of falling string figure 4 . The only way to determine the energy loss of a jet precisely in strongly coupled regime is solving the gravitational bulk-to-boundary problem. One can solve Einstein's equations for the perturbation in the 5d geometry due to the presence of the string and according to the bulk to boundary map interpret the near boundary behavior of the metric perturbation as the perturbation in the SYM energy-momentum tensor by the presence of jet which will be left for the future work.

References

[1] Adams J *et al.* (STAR Collaboration) 2005 *Nucl.Phys.* **A757** 102–183 (*Preprint* nucl-ex/0501009)
[2] Adcox K *et al.* (PHENIX Collaboration) 2005 *Nucl.Phys.* **A757** 184–283 (*Preprint* nucl-ex/0410003)
[3] Arsene I *et al.* (BRAHMS Collaboration) 2005 *Nucl.Phys.* **A757** 1–27 (*Preprint* nucl-ex/0410020)
[4] Back B, Baker M, Ballintijn M, Barton D, Becker B *et al.* 2005 *Nucl.Phys.* **A757** 28–101 (*Preprint* nucl-ex/0410022)
[5] Policastro G, Son D T and Starinets A O 2001 *Phys.Rev.Lett.* **87** 081601 (*Preprint* hep-th/0104066)
[6] Kovtun P, Son D T and Starinets A O 2005 *Phys.Rev.Lett.* **94** 111601 (*Preprint* hep-th/0405231)
[7] Maldacena J M 1999 *Int.J.Theor.Phys.* **38** 1113–1133 (*Preprint* hep-th/9711200)
[8] Witten E 1998 *Adv.Theor.Math.Phys.* **2** 253–291 (*Preprint* hep-th/9802150)
[9] Gubser S, Klebanov I R and Polyakov A M 1998 *Phys.Lett.* **B428** 105–114 (*Preprint* hep-th/9802109)
[10] Aharony O, Gubser S S, Maldacena J M, Ooguri H and Oz Y 2000 *Phys.Rept.* **323** 183–386 (*Preprint* hep-th/9905111)
[11] Casalderrey-Solana J, Liu H, Mateos D, Rajagopal K and Wiedemann U A 2011 (*Preprint* 1101.0618)
[12] Yin Z B (ALICE Collaboration) 2013 *Acta Phys.Polon.Supp.* **6** 479–484
[13] Aad G *et al.* (ATLAS Collaboration) 2010 *Phys.Rev.Lett.* **105** 252303 (*Preprint* 1011.6182)
[14] Chatrchyan S *et al.* (CMS Collaboration) 2011 *Phys.Rev.* **C84** 024906 (*Preprint* 1102.1957)
[15] Collaboration C (CMS Collaboration) 2012
[16] Karch A and Katz E 2002 *JHEP* **0206** 043 (*Preprint* hep-th/0205236)
[17] Herzog C, Karch A, Kovtun P, Kozcaz C and Yaffe L 2006 *JHEP* **0607** 013 (*Preprint* hep-th/0605158)
[18] Chesler P M, Jensen K, Karch A and Yaffe L G 2009 *Phys.Rev.* **D79** 125015 (*Preprint* 0810.1985)
[19] Janik R A and Peschanski R B 2006 *Phys.Rev.* **D73** 045013 (*Preprint* hep-th/0512162)
[20] Horowitz W A 2010 (*Preprint* 1011.4316)

Review of jet reconstruction algorithms

Ryan Atkin

School of Chemistry and Physics, University of Kwa-Zulu Natal, Westville campus, Private Bag X54001, Durban 4000, South Africa

E-mail: rjatkin93@gmail.com

Abstract. Accurate jet reconstruction is necessary for understanding the link between the unobserved partons and the jets of observed collimated colourless particles the partons hadronise into. Understanding this link sheds light on the properties of these partons. A review of various common jet algorithms is presented, namely the K_t, Anti-K_t, Cambridge/Aachen, Iterative cones and the SIScone, highlighting their strengths and weaknesses. If one is interested in studying jets, the Anti-K_t algorithm is the best choice, however if ones interest is in the jet substructures then the Cambridge/Aachen algorithm would be the best option.

1. Introduction

A jet can be defined as a collimated spray of stable particles arising from the fragmentation and hadronisation of a parton(quark or gluon) after a collision. Jet reconstruction algorithms are used to combine the calorimetry and tracking information to define jets. The jets provide a link between the observed colourless stable particles and the underlying physics at the partonic level. A basic illustration of a collision of two protons, the subsequent particle shower and a reconstructed jet is shown in Fig.1. This link provides information on the kinematics of the originating partons which can be used to shed light on quantum chromodynamics(QCD) and infer the presence and thus the properties of the Higgs boson and other particles too short lived to be detected. An accurate jet algorithm will also be able to calculate the correct amount of missing energy in the detector, which could allude to new physics beyond the standard model if found not to match theory[1].

2. Important Aspects

Some aspects of an algorithm that need to be considered are the jet size and whether the algorithm is infra-red and collinear(IRC) safe. The jet size and area determine the susceptibility of a jet to soft radiation. A larger jet radius is important as it allows the jet to capture enough of the hadronised particles for the accurate calculation of the jets mass and energy. However a smaller jet radius is useful in reducing the amount of the underlying event(UE) and pile-up(PU) captured by the jet, preventing the overestimation of the jets mass and energy.

The splitting of a hard particle, while using a collinear unsafe algorithm, will result in the altering of the number and contents of the jets. A similar problem arises when a soft gluon is added to the system while an infra-red unsafe algorithm is in use. An IRC unsafe algorithm will affect the perturbative QCD calculations.

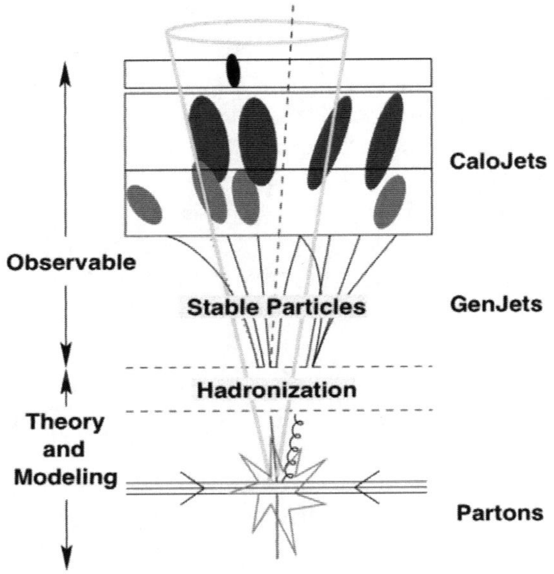

Figure 1. A simple example of an event showing the point of collision, the fragmentation and hadronization of the quarks and gluons and the resulting jet found through the detection of the stable particles[2]. Calojets are those jets created using the calorimeter output whereas Genjets are jets created using stable simulated particles. The dashed line represents the direction of the missing energy.

3. Jet Algorithms

There are two main classes of jet algorithms in use. The first being the cone algorithms, of which the most important are the iterative cone with progressive removal(IC-PR)(iterative version of [3]), the iterative cone with split-merge procedure(IC-SM)[4] and the seedless infra-red safe cone(SIScone)[5]. The second class is the sequential clustering algorithms which comprises of the K_t[6], Anti-K_t[7] and the Cambridge/Aachen[8] algorithms.

3.1. Cone Algorithms

Cone algorithms assume that particles in jets will show up in conical regions and thus they cluster based on (η-ϕ) space, resulting in jets with rigid circular boundaries. Cone algorithms in the past were preferred by experimentalists as they were easier to implement, however they are not favoured by theorists as they contain non-physical constants. Cone algorithms are generally IRC unsafe as well.

3.1.1. IC-PR: The iterative cone algorithm with progressive removal is a collinear unsafe algorithm. The CMS iterative cone and Pythia cone are examples of IC-PR. Its method of clustering is as follows:

Find the hardest(largest p_t) cell and make it a seed. Create a cone of radius R around this seed and calculate the trial jet axis by summing up the cells within this cone using four-vectors. If the trial jet axis is equal to the seed axis, the cone is labelled as stable and all the particles within the stable cone are removed from the list of particles. The next hardest remaining cell is found and this procedure is repeated. But if the trial jet axis does not equal the seed axis, the trial jet axis is made the new seed axis and the process is repeated until convergence of the

Journal of Physics: Conference Series **645** (2015) 012008 doi:10.1088/1742-6596/645/1/012008

axes occurs. The entire process is repeated until there are no seeds left above a threshold energy E_{cut}.

3.1.2. IC-SM: The iterative cone algorithm with the split merge procedure is an infra-red unsafe algorithm. Examples include JetClu, midpoint cone and the ATLAS cone. Its method of jet formation is as follows:
First make all cells above a threshold energy E_{cut} a seed. Find all stable cones associated with those seeds using the same process as in the IC-PR algorithm but without removing any particles from the list once a stable cone is found. All stable cones found are labelled as protojets and a split merge procedure is run on the protojets once they have all been found. The split merge procedure is described in Appendix A.

3.1.3. SIScone: The seedless infra-red safe cone algorithm is the only IRC safe cone algorithm. Due to it's relatively smaller area, it is not badly affected by the UE and PU. It also has the ability to reach further than R for hard radiation and thus has good resolution, however this fact means that it is bad at resolving multijets. The following is the SIScone's method of jet finding:

For particle i
  Find all particles j within distance $2R$ of i
  If there is no particle j
    i is labelled a stable cone and added to the list of protojets
  Else
    Find both circles created by i and j lying on their circumference and calculate the momenta of the cones they define.
    For each circle
      Consider all four permutations of the two edge points being contained in or out of the circle. These four circles are labelled as current cones.
      For each current cone not previously found
        Determine whether the current cone's in or out status of the edge particles is the same as the cone defined by the momentum of the particles within the aforementioned current cone. If not, label the current cone as unstable.
Perform an explicit stability check on all current cones not labelled as unstable and add all stable cones to the list of protojets.
Run a split merge procedure on the protojets.

3.2. Sequential Clustering Algorithms
Sequential clustering algorithms assume that particles within jets will have small differences in transverse momenta and thus groups particles based on momentum space, resulting in jets that have fluctuating areas in (η-ϕ) space. Sequential clustering algorithms have always been favoured by theorists but not by experimentalists as in the past they had slow computational performance. However since the introduction of the FastJet program[9], clustering algorithms are much faster and preferred by experimentalists as well. Sequential clustering algorithms are also IRC safe.
All sequential clustering algorithms have a similar method. The first distance variable is the one between two particles $d_{ij} = min(p_{ti}^a, p_{tj}^a) \times \frac{R_{ij}^2}{R}$, where a is an exponent corresponding to a particular clustering algorithm, $R_{ij}^2 = (\eta_i - \eta_j)^2 + (\phi_i - \phi_j)^2$ is the (η-ϕ) space distance between the two particles and R is the radius parameter which determines the final size of the jet and is usually between 0.4 - 0.7. The second distance variable is $d_{iB} = p_{ti}^a$ and is the momentum space distance between the beam axis and the detected particle.

The sequential clustering algorithms work by first finding the minimum of the entire set $\{d_{ij}, d_{iB}\}$. If d_{ij} is the minimum then particles i and j are combined into one particle (ij) using summation of four-vectors after which i and j are removed from the list of particles. If d_{iB} is the minimum, i is labelled a final jet and removed from the list of particles. This process is repeated until either all particles are part of a jet with the distance between the jet axes R_{ij} greater than R, which is inclusive clustering. Or until a desired amount of jets have been found, this is exclusive clustering.

3.2.1. K_t*:* The a value corresponding to the K_t algorithm is 2, resulting in the following equations:

$$d_{ij} = min(p_{ti}^2, p_{tj}^2) \times \frac{R_{ij}^2}{R} \tag{1}$$

$$d_{iB} = p_{ti}^2 \tag{2}$$

The dominance of low p_t is shown in Eq. (1) and so the K_t algorithm prefers to cluster soft particles first, resulting in an area that fluctuates considerably and an algorithm that is susceptible to the UE and PU. Due to its method of clustering, K_t does a good job at resolving subjets.

3.2.2. Anti-K_t*:* The a value corresponding to the Anti-K_t algorithm is -2, resulting in the following equations:

$$d_{ij} = min\left(\frac{1}{p_{ti}^2}, \frac{1}{p_{tj}^2}\right) \times \frac{R_{ij}^2}{R} \tag{3}$$

$$d_{iB} = \frac{1}{p_{ti}^2} \tag{4}$$

So Eq. (3) is dominated by high p_t and the algorithm prefers to cluster hard particles first. Thus the area only fluctuates slightly and the algorithm is only slightly susceptible to the UE and PU. The Anti-K_t's clustering preference results in a algorithm that is the best at resolving jets but due to its poor de-clustering, it is the worst for studying jet substructure.

3.2.3. Cambridge/Aachen: The a value corresponding to the C/A algorithm is 0, resulting in the following equations:

$$d_{ij} = \frac{R_{ij}^2}{R} \tag{5}$$

$$d_{iB} = 1 \tag{6}$$

Both of the distance variables are independent of momentum and so its area fluctuates somewhat and is somewhat susceptible to the UE and PU. An example of the four main algorithms' jet areas are illustrated in Fig. 2. Due to the purely spatial character of the distance variables, C/A de-clusters the best and so is the best suited for studying jet substructure, although it is slightly more complicated to de-cluster than the K_t algorithm.

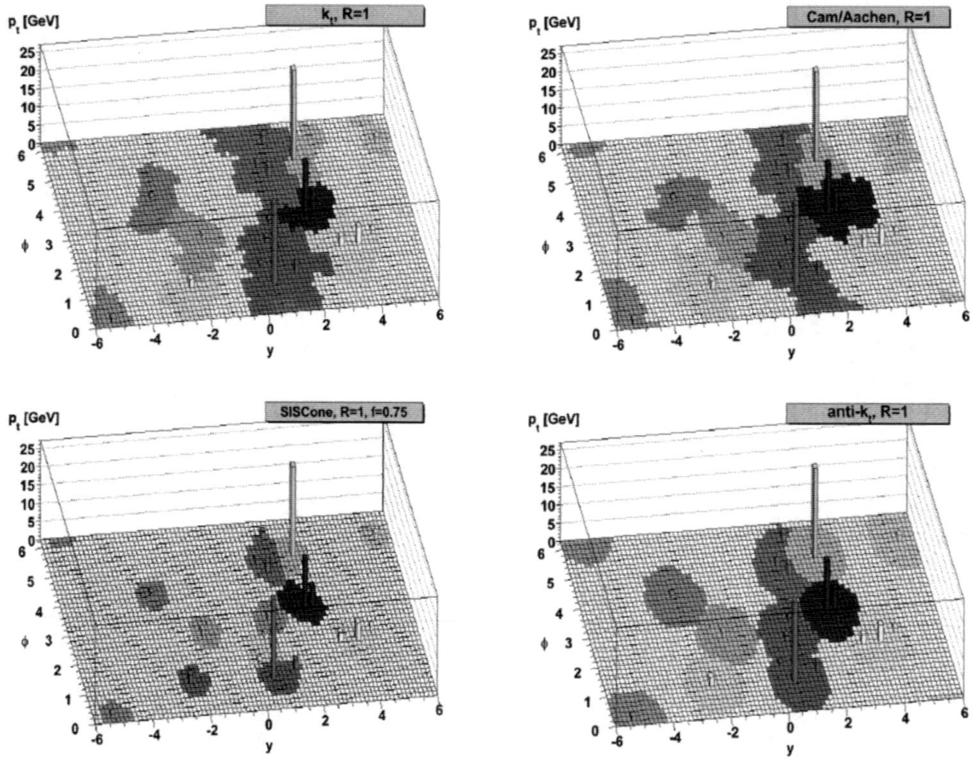

Figure 2. The four main jet reconstruction algorithms' areas, performed on the same data with the same input radius[7]. Noted features are the high irregularity in the the K_t algorithms area, the conical shape of the Anti-K_t's jets illustrating this algorithms preference for hard radiation and the smaller effective radius of the SIScone, due to the split merge procedure, which can be observed via smaller jet areas and two jets being resolved in the place of just the one grey jet. The different colours are used to represent the different jets and their areas.

4. Conclusion

When using jet reconstruction algorithms, two important things to consider are infra-red and collinear safety. Due too this the iterative cone with split merge procedure and the iterative cone with progressive removal are no longer widely used, as they are not both infra-red and collinear safe. Other important aspects to consider are jet size and shape. A large enough jet is necessary to capture the required amount of non-perturbatively hadronised particles while not being too large as to capture an excessive amount of the underlying event and pile-up.
The most accurate jet algorithm for resolving jets is the Anti-K_t algorithm whereas for the study of jet substructure, the Cambridge/Aachen algorithm is best suited. Accurate jet reconstruction is the only way we are able to study the properties of quarks and gluons as they are unable to be directly observed.

Acknowledgments
I would like to thank my supervisor Dr. Sahal Yacoob for his important input and assistance, Prof. Bruce Mellado and Prof. Alan Cornell of the University of the Witwatersrand for organising the High Energy Particle Physics workshop held at iThemba labs in Gauteng and finally the NRF for funding my accommodation and transport during my stay at the workshop.

Appendix A - Split merge procedure
The following is the method used in the split merge procedure[5]:
Remove all protojets with $p_t < p_{t,cut}$ (Helps reduce effects of pile-up)
Repeat
  Find the hardest protojet i through scalar sum of constituents' p_t
    Find the next hardest protojet j that shares particles with i
    If there is no overlapping protojet j
      i is labelled a final jet and removed from the list of particles
    Else
      If $p_{t,shared} < f \times p_{tj}$
        Share the particles according to which axis they are closer to and recalculate the protojets' momenta
      Else
        Merge protojets i and j into one and remove i and j from the list of protojets
      If either of these two processes produces another protojet matching one already found, keep both as two distinct protojets
Until (no protojets left)

The merging parameter f is usually between the values of 0.5 and 0.75 and determines the size of the jets after the split merge procedure.

References
[1] M. Hodgkinson, in Proceedings of the 34th International Conference in High Energy Physics (ICHEP08), Philadelphia, 2008, eConf C080730 [arXiv:hep-ex/0810.0181]
[2] P. Schieferdecker. Available at: <https://twiki.cern.ch/twiki/bin/viewfile/Sandbox/Lecture?rev=1;filename =Philipp_Schieferdeckers_Lecture.pdf> [Accessed 17 December 2015]
[3] G. Arnison et al. [UA1 Collaboration], Phys. Lett. B 132 (1983) 214.
[4] G. C. Blazey, J. R. Dittmann, S. D. Ellis, V. D. Elvira, K. Frame, S. Grinstein, R. Hirosky and R. Piegaia *et al.* [arXiv:hep-ex/0005012]
[5] G. P. Salam and G. Soyez, JHEP **0705** (2007) 086 [arXiv:0704.0292 [hep-ph]].
[6] S. Catani, Y.L. Dokshitzer, M.H. Seymour, and B.R. Webber. Longitudinally invariant Kt clustering algorithms for hadron hadron collisions. Nucl.Phys., B406:187224, 1993.
[7] M. Cacciari, G.P. Salam, and G. Soyez. JHEP **0804** (2008) 063 [arXiv:0802.1189 [hep-ph]].
[8] CMS collaboration. A Cambridge-Aachen(C-A) based jet algorithm for boosted top-jet tagging. CMS PAS JME-09-001, 2009
[9] M. Cacciari, G.P. Salam, and G. Soyez, Eur.Phys.J. C72 (2012) 1896 [arXiv:1111.6097 [hep-ph]]

Understanding the Higgs boson with the Large Hadron Electron Collider

Jonathan Esteves

School of Physics, University of the Witwatersrand, Johannesburg 2050, South Africa.

E-mail: jonathan.esteves@students.wits.ac.za

Abstract. The Large Hadron Electron Collider (LHeC) at the European Laboratory, CERN, is expected to collide electrons and protons at high energy. Studies pertaining to the feasibility of observing the Higgs boson in this environment were reported in the Conceptual Design Report. Here the effect of decreasing the electron energy in an ep collision to find the optimal, economic electron energy for the study of the Higgs boson in the future LHeC is studied. Two production mechanisms are addressed: a Higgs boson production process in an ep collision, and a background process. The electron energy was varied between 10 GeV and 100 GeV in increments of 10 GeV. The results obtained in this study have shown that using an electron energies between 40 GeV and 60 GeV would be sufficient to measure the properties of the Higgs boson without compromising on the validity of obtained results.

1. Introduction and Theory

Peter Higgs proposed a mechanism that allowed for the existence of a fundamental particle, the higgs boson, almost half a century (48 years) before its discovery at CERN's LHC (Large Hadron Collider), in Switzerland on July 4th 2012 [1][2].

On October 8th 2013, the higgs boson discovery won Francois Englert and Peter Higgs the Nobel prize in physics, and resulted in: better understanding the origin of mass, confirmation of theoretical assumptions made with respect to the Standard Model, validation of theory related to electroweak forces and supersymmetry, and lastly it opens a new sector to study. The higgs production process has not yet been studied to its potential, and fortunately requires a relatively cost-effective, accelerator complex.

The current electron energy estimatation of 60 GeV is challeneged by observing the effect a decrease of electron energy has on the production of the higgs boson, in an electron-proton collision. CERN, a European Laboratory, are planning on constructing the LHeC (Large Hadron Electron Collider) such that accelerated electrons from the LHeC can collide with accelerated protons from the LHC [3]. Studies pertaining to the feasibility of observing the higgs boson in this environment, has been conducted by generating Monte Carlo events (50 000 events were generated) for two processes at different electron energies, differing by 10 GeV in a [10, 100] GeV interval.

Signal, the first process studied, is described by an electron-proton collision resulting in an electron neutrino, a higgs boson, and a jet. Figure 1 depicts the most probable, lowest-order

Feynman diagrams for single higgs production in an electron -proton collision [4]. The background process is described by an electron-proton collision resulting in an electron neutrino, bottom quark, bottom anti-quark and a jet. Figure 2 depict four of the lowest-order Feynman diagrams for the background collisions considered [4].

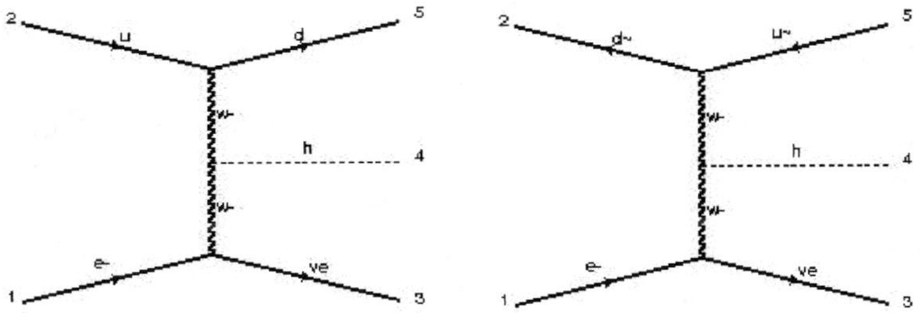

Figure 1. Two lowest-order Feynman diagrams for single higgs production in an electron - proton collision

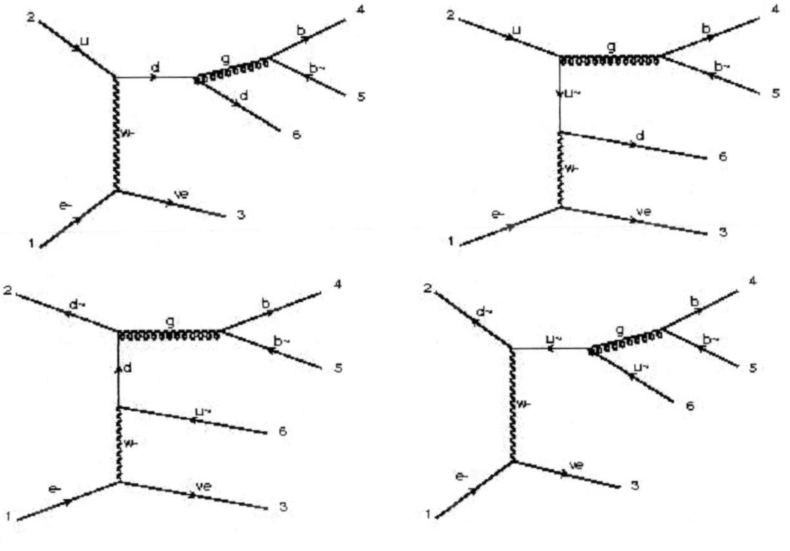

Figure 2. Four lowest-order Feynman diagrams of background collisions considered

In the standard model, gauge bosons mediate the forces and are associated with the $SU(3)_{colour}$ x $SU(2)_L$ x $SU(1)_Y$ group symmetries. The gluonic force of the strong interactions is described by the $SU(3)_{colour}$ factor whilst the unified electroweak interactions are described by the $SU(2)_L$ x $SU(1)_Y$ factors [5].

Journal of Physics: Conference Series **645** (2015) 012009 doi:10.1088/1742-6596/645/1/012009

The invariance of the $SU(2)_L$ x $SU(1)_Y$ factors are broken down to the subgroup denoted by $U(1)_{EM}$ (the unbroken gauge invariance of electromagnetism subgroup). This results in four gauge bosons namely W^-, W^+, Z and the γ (photon) [5][6].

This symmetry is not necessarily broken by explicit interaction of the subatomic particles, but rather they are broken by the asymmetry of the state of lowest energy. In quantum field theory, the framework of spontaneous symmetry breaking and the symmetry of the state of lowest energy is known as a vacuum [6].

Each time a spontaneously broken direction (which gives rise to the Goldstone boson) corresponds to a gauge symmetry, the associated Goldstone boson and a massless gauge boson combine to form a massive gauge boson. This is known as the higgs mechanism [6]. This is the same mechanism that was proposed by Peter Higgs in the 1960's, which hypothesizes that there must exist some sort of 'lattice' that fills the universe. This lattice is formally known as the Higgs field [7].

The Higgs field is defined as; "hypothetical physical field that endows elementary particles with mass and that is mediated by the Higgs boson". There are many analogies for the Higgs field but the most popular is of a famous person walking into a room that is relatively densely occupied with people. As the famous person enters the room, the people closest to the entrance are drawn to him/her. The famous person attracts people that are close by as he/she moves through the room. The clustering at each of the famous person's steps can be seen as an increase in resistance to movement, i.e. the famous person has gained mass. If a normal (non-famous) person was to enter the room, no one would be drawn to him/her and they would not have gained any mass. In this same respect, not all particles affect the field (they are massless). A particle with mass would experience greater 'resistance'to change in the presence of the Higgs field and would require energy to commence movement once it has been stopped [7].

One of the optimal conditions for the study of the Higgs boson is the minimisation of background detected in the same region of the detector as signal detection. This conceptual report studies the effect on the following predicted properties pertaining to the processes under consideration: cross section (measure of area in collision process), rapidity (measure of rate of motion in the induced electromagnetic field), transverse momentum (momentum measurement along beam axis), distance in the rapidity-phi plane (distance between particles of interest) and invariant mass (mass measurement in an invariant basis). Studying these predicted properties will allow for further insight into the processes system behaviour, as function of electron energy.

2. Results

When generating data for two electron - proton collision processes (constant proton energy of 7000 GeV and varied electron energy), a total of 300 kinematic distributions were obtained. This data was processed and the electron dependence of the average; cross section, rapidity, transverse momentum, distance in the (rapidity, phi) plane and invariant mass, is illustrated in figures 3 to 11.

Journal of Physics: Conference Series **645** (2015) 012009 doi:10.1088/1742-6596/645/1/012009

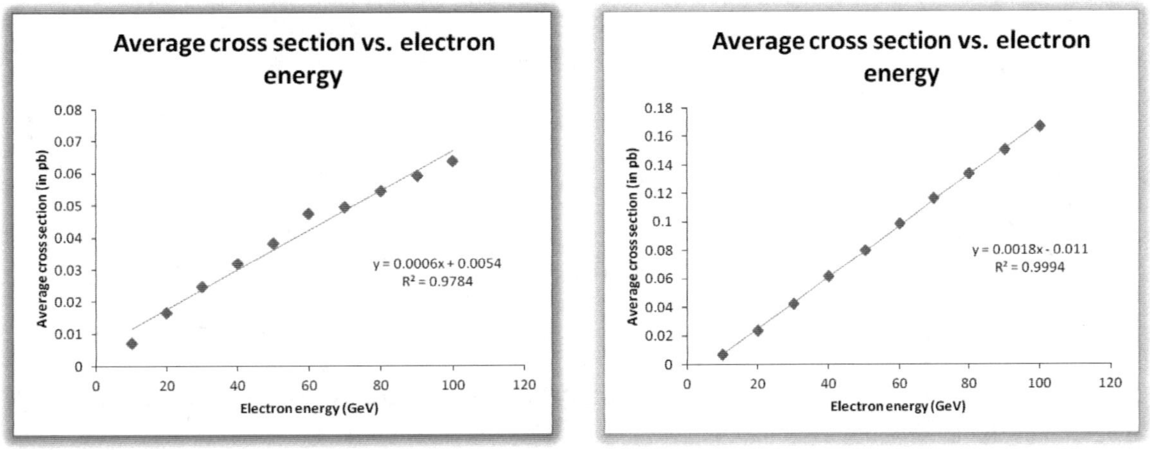

Figure 3. The average cross section as a function of electron energy for the Higgs boson production process and the background process, left and right respectively.

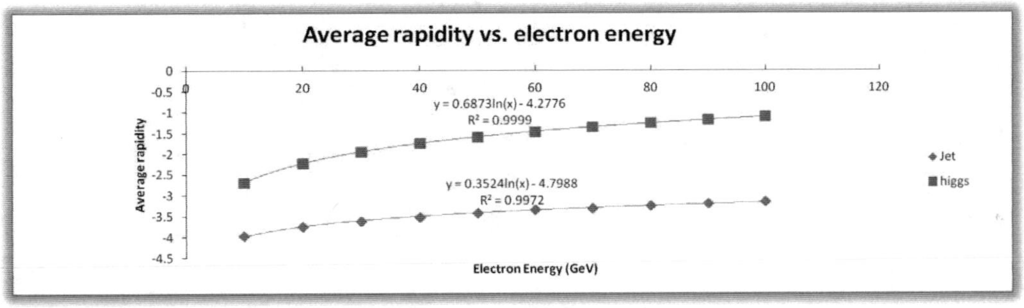

Figure 4. The average rapidity as a function of electron energy for the Higgs boson production process.

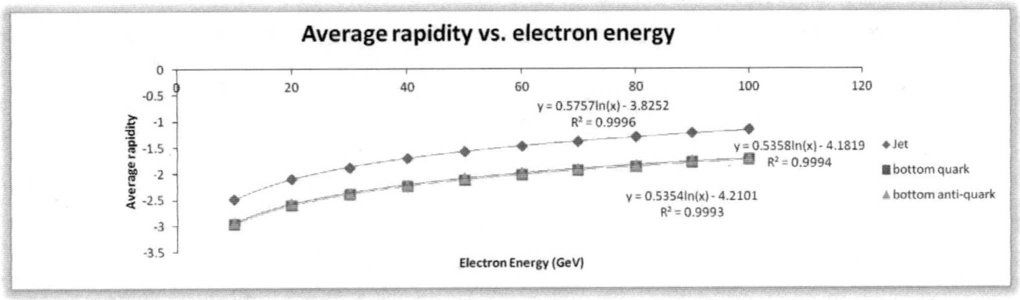

Figure 5. The average rapidity as a function of electron energy for the background process.

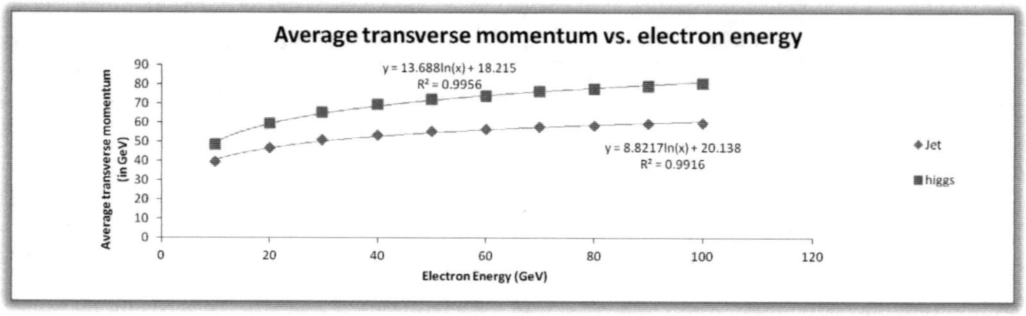

Figure 6. The average transverse momentum as a function of electron energy for the Higgs boson production process.

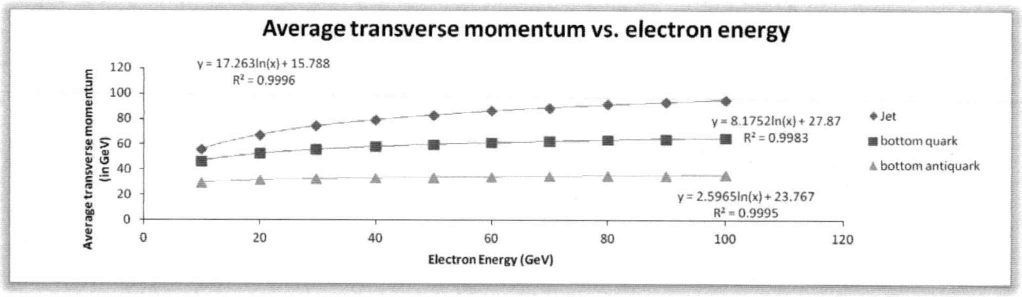

Figure 7. The average transverse momentum as a function of electron energy for the background process.

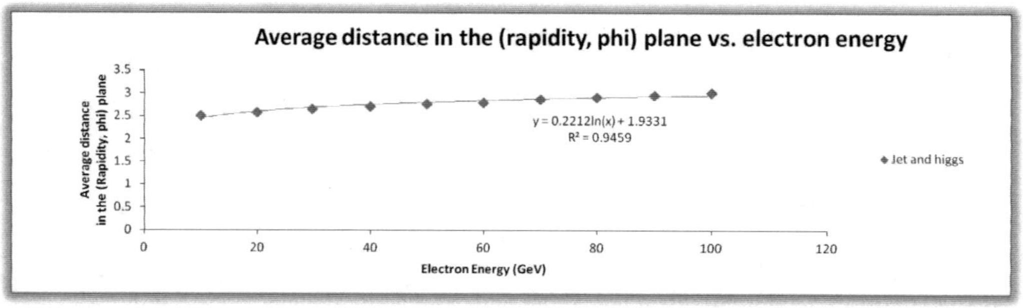

Figure 8. The average distance in the (rapidity, phi) plane as a function of electron energy for the Higgs boson production process.

Journal of Physics: Conference Series **645** (2015) 012009 doi:10.1088/1742-6596/645/1/012009

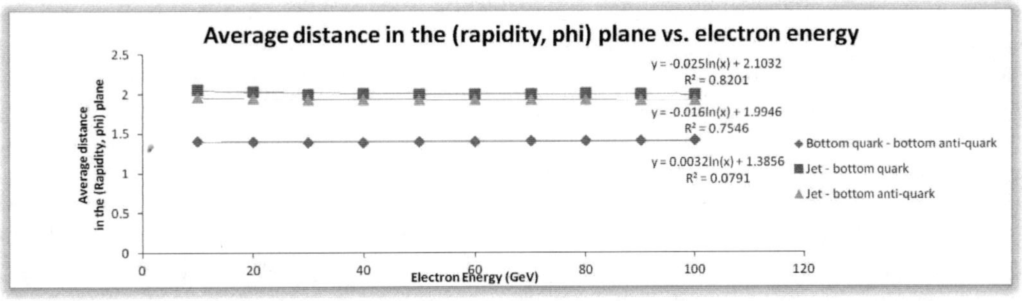

Figure 9. The average distance in the (rapidity, phi) plane as a function of electron energy for the background process.

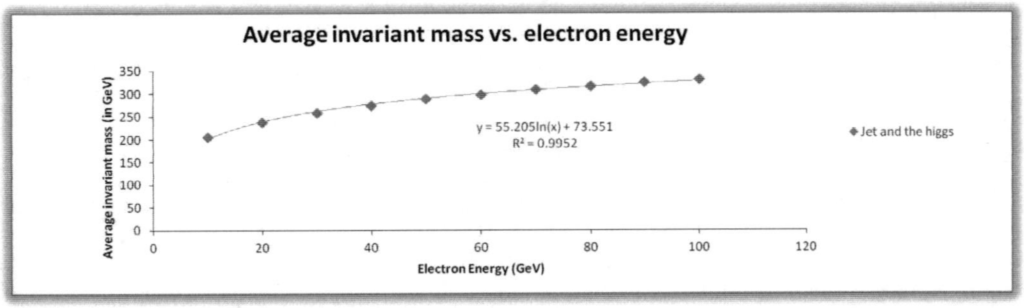

Figure 10. The average invariant mass as a function of electron energy for the Higgs boson production process.

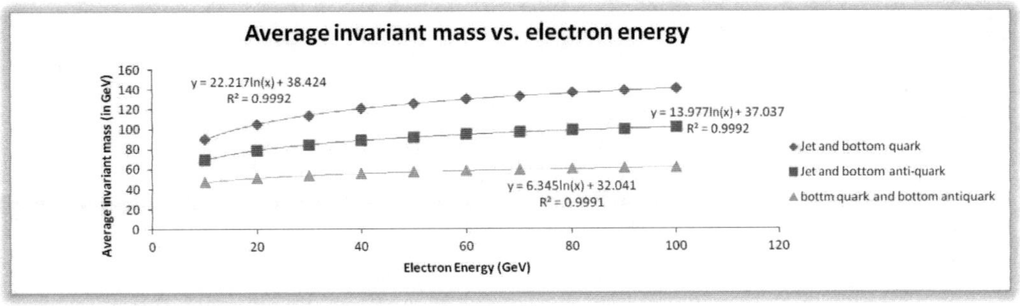

Figure 11. The average invariant mass as a function of electron energy for the background process.

3. Discussion and Conclusion

The relationship between cross section and electron energy, in both signal and background, is best described by linear growth [refer to figure 3]. The physical interpretation of this is; there is more energy and therefore more particle taking place in the collision, i.e. a larger cross section is obtained with an increase in electron energy. A difference in gradient is noted between signal and background, depicting characteristic behaviour of the two systems. Studying the rapidity to electron dependence reveals that as one decreases the electron energy, the Higgs rapidity decreases (becoming more comparable to the rapidity of the jet in background) [refer to figures 4 and 5]. The Higgs rapidity is expected to be indistinguishable from the background rapidity for electron beam energies lower than 40 GeV. This is physically justified by the characteristic rapidity properties of the Higgs boson production process; the jet in the Higgs process has lower rapidity than that of the higgs boson whilst in background; the scattered quarks have a lower rapidity than the jet. The Higgs rapidity is greater than that of the jet in background, however the Higgs system is camouflaged by background as a lower electron beam energy limit is taken. Rapidity considerations has resulted in the placement of a lower bound on electron beam energy, 40 GeV. Considerations of transverse momentum, distance in the (rapidity, phi) plane and invariant mass revealed weak electron beam energy dependence [refer to figures 6, 7, 8, 9, 10 and 11].

Results have shown that using an electron energy between 40 GeV and 60 GeV would be sufficient to measure properties of the Higgs boson, without compromising on the validity of obtained results. This conclusion is founded mainly on rapidity to electron dependence of the two considered processes in an electron-proton collision.

Acknowledgements
Prof. B. Mellado, The University of the Witwatersrand School of Physics.

References
[1] Aad G et al [ATLAS Collaboration] 2012 Phys. Lett. B 716 1.
[2] Chatrchyan S et al [CMS Collaboration] 2012 Phys. Lett. B 716 30.
[3] T. Han and B. Mellado, Phys.Rev. D82 (2010) 016009.
[4] Mad Graph, madgraph.hep.uiuc.edu.
[5] Hyper Physics, hyperphysics.phy-satr.gsu.edu/hbase/forces/unify.html.
[6] AR Journals, arjournals.annualreviews.org.
[7] Exploratorium , www.exploratorium.edu/origins/cern/ideas/higgs.html.

Measurement of the Higgs Boson Transverse Momentum and its Sensitivity to New Physics Beyond the Standard Model

D Gossman[1]
[1] Physics Building, University of the Witwatersrand, 1 Jan Smuts Avenue, Braamfontein, Johannesburg, South Africa

E-mail: dmgossman@gmail.com

Abstract. When simulations based on the Standard Model (SM) of particle physics are compared to actual data obtained by the ATLAS and CMS experiments at the CERN Large Hadron Collider (LHC). It is seen that there is an excess in the transverse momentum cross-sectional data above what is predicted by the simulations. In order to make predictions for higher centre of mass energies at the LHC, simulations of processes resulting in the production of Higgs bosons were done for different centre of mass energies. At the energy scales seen at the LHC, the SM predicts that the main production mechanism for Higgs bosons is gluon fusion. The production of a Higgs boson in this manner must be accompanied by the production of one or more other particles in order for the Higgs boson to acquire transverse momentum (P_T). This is because there is no transvers momentum coming into the collision and so conservation of momentum requires two or more particles with opposite P_T to be produced in order for them to have non-zero P_T. If a heavy scalar boson is produced in this interaction which decays into a Higgs boson and some other particle, the emission of this other particle would give the Higgs boson extra transverse momentum above what is predicted by the SM.

1. Theoretical background

1.1. The Standard Model of particle physics

The standard model of particle physics (SM) is a theory that stems from quantum field theory (QFT) and explains three of the four fundamental forces as interactions between the fundamental particles. These forces are the electromagnetic, the weak nuclear and the strong nuclear forces. In the fundamental particles of SM there are 12 matter particles separated into three generations and further separated into 6 quarks and 6 leptons, as well as 4 gauge bosons and a scalar boson (the Higgs). (Figure 1)

Journal of Physics: Conference Series **645** (2015) 012010 doi:10.1088/1742-6596/645/1/012010

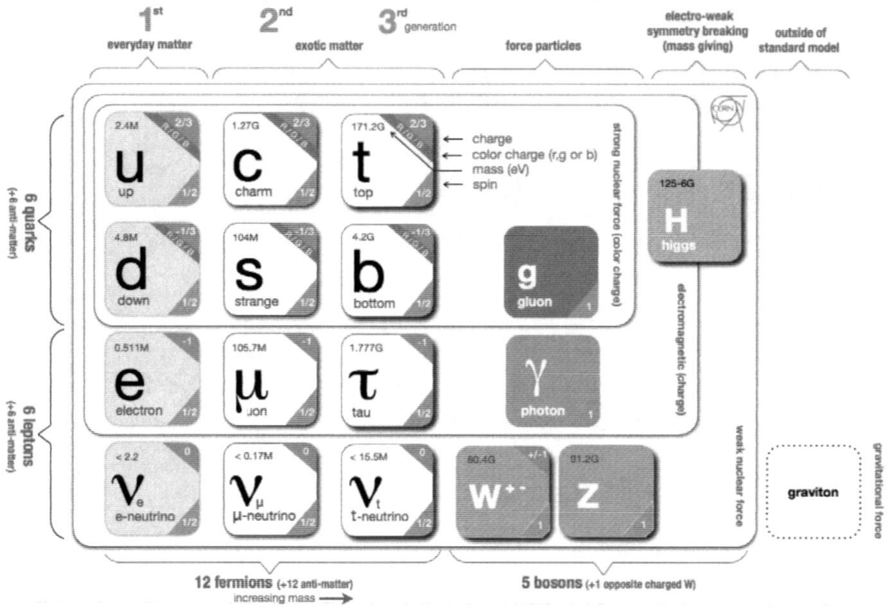

Figure 1: The SM infographic

The three generations of matter particles are called fermions and all have a spin of 1/2. The gauge bosons on the other hand, have a spin of 1 and the Higgs boson has zero spin. The gauge bosons are considered the force carriers and although this is a foreign concept the forces mentioned above are understood to be experienced by the transfer of virtual gauge bosons. The gluon is responsible for the strong nuclear force and only interacts with quarks and itself. The photon is responsible for the electromagnetic force and only interacts with charged particles, and the W and Z bosons are responsible for the weak force and interact with all matter particles. The Higgs is responsible for giving mass to the massive particles, the quarks, leptons and W and Z bosons. [2]

1.1.1. Gauge invariance. In gauge theory the redundant degrees of freedom in a Lagrangian are referred to as gauges. A symmetry group (such as SU(3) in QCD) then defines the transformations between these gauges. Associated with the symmetry group are group generators for which there necessarily arise vector fields or gauge fields for each group generator, which must be added to the Lagrangian in order to ensure its gauge invariance. The quantization of these vector fields result in what are gauge bosons. For example SU(3) has 8 group generators which result in the 8 gluons in QCD. Now a form of the SM Lagrangian which obeys SU(3)×SU(2)×U(1) gauge invariance can be written down in which all the particles have no mass which is a problem because there is a vast amount of experimental data saying fermions and W and Z bosons have mass and furthermore if we simply give these particles mass the gauge invariance of SU(2)×U(1) is violated. Thus, a mechanism that gives mass without violating gauge invariance is needed. This mechanism in spontaneous electroweak symmetry breaking.

1.1.2. Spontaneous electroweak symmetry breaking. In order to reconcile SU(2)×U(1) symmetries another interaction must be introduced, which turns out to be the Higgs mechanism. This is done by introducing a potential V(φ) dependent on a scalar field φ that is required to be symmetric about $\varphi \longrightarrow -\varphi$:

$$V(\varphi) = \mu^2 \varphi^\dagger \varphi - \lambda(\varphi^\dagger \varphi)^2 \quad (2)$$

By requiring $\lambda > 0$ and $\mu^2 < 0$, $V(\varphi)$ will have two minima at $\varphi = \pm v = \pm\sqrt{-\mu^2/\lambda}$. The minima are removed from $\varphi = 0$ and thus the system will spontaneously break symmetry when it moves into a ground state. The Lagrangian for this potential is:

$$\mathcal{L}_\varphi = (D^\mu \varphi)^\dagger (D_\mu \varphi) - V(\varphi) = (D^\mu \varphi)^\dagger (D_\mu \varphi) - \frac{1}{2}\mu^2 \varphi^2 - \frac{1}{4}\lambda \varphi^4 \quad (1)$$

Where D_μ is the covariant derivative and $\mathcal{L}_\varphi$ forms the scalar component of the SM Lagrangian Now in order to give the $W^\pm$ and Z bosons mass while leaving the photon massless the scalar field φ is defined as a complex doublet:

$$\varphi = \begin{pmatrix} \varphi^+ \\ \varphi^- \end{pmatrix} = \frac{1}{\sqrt{2}} \begin{pmatrix} \varphi_1 - i\varphi_2 \\ \varphi_3 - i\varphi_4 \end{pmatrix} \quad (2)$$

$$\Rightarrow V(\varphi) = \mu^2 \varphi^\dagger \varphi - \lambda(\varphi^\dagger \varphi)^2 \quad (3)$$

If it is chosen that $\varphi_1 = \varphi_2 = \varphi_4 = 0$ and $\varphi_3 = \frac{1}{\sqrt{2}}\left(-\frac{\mu^2}{\lambda}\right)^{\frac{1}{2}}$ then it can be shown that:

$$\varphi = \frac{1}{\sqrt{2}} \begin{pmatrix} 0 \\ v + h \end{pmatrix} \quad (4)$$

Where h is the real component of the neutral scalar field φ. These corrections as well as the necessary Yukawa terms then make up the SU(2)xU(1) Lagrangian which can be used to calculate the strength of Higgs to other particle couplings.

1.1.3. Quantum field theory and the SM. The mathematics of the SM comes from quantum field theory (QFT) in which the partials of the SM are considered fields that permeate space-time and the interactions between the fields are understood by Lagrangian mechanics. The internal symmetries of the system are usually defined, and then the most general renormalizable Lagrangian that obeys these symmetries can be defined. There are three fundamental interactions; the strong, the weak and the electromagnetic. The electromagnetic and weak interactions can however be combined into one interaction called the electroweak interaction. The gauge symmetry of SM is SU(3)×SU(2)×U(1) where each of the factors are said to give rise to the three interactions. [2] SU(3) for Quantum Chromo Dynamics (QCD), the strong interaction, SU(2) for the weak interaction and U(1) for the electromagnetic interaction. The SM Lagrangian for these symmetries can be written in terms of the QCD sector and the EW sector:

$$\mathcal{L}_{SM} = \mathcal{L}_{SU(3)} + \mathcal{L}_{SU(2)\times U(1)} \quad (1)$$

1.2. Higgs production at the LHC

At the CERN Large Hadron Collider (LHC) protons are accelerated to speeds close to the speed of light in opposite directions and then smashed together. [3] This process is energy intensive enough to cause the protons to split up into its constituents. These constituents then interact in interesting ways that reveal the nature of the forces governing these interactions. One such interaction isHiggsproduction through gluon fusion via an intermediate top quark loop ($gg \longrightarrow H$): [4]

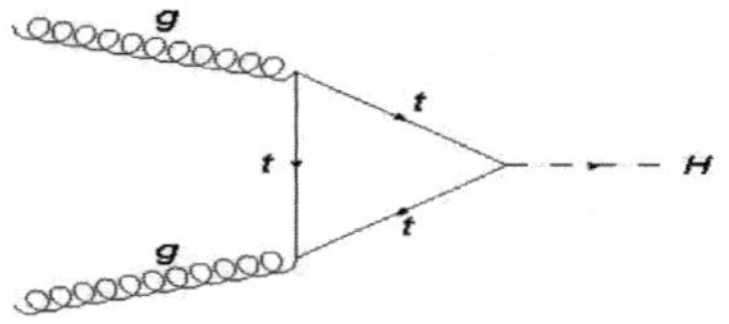

Figure 2: Higgs production via gluon-gluon fusion [5]

This is in fact the leading production mechanism of the many partonic processes producing theHiggsat the LHC. [4] The problem with this interaction is that the Higgs must have transvers momentum (P_T) in order to be observed. So one needs to look at the case where one or more other particles are produced as well as the Higgs. These particles decay into hadronic jets and because the net P_T must be zero the Higgs will have the same P_T in the opposite direction as the net P_T of the jets. One such example is ($gg \longrightarrow gH$): [4]

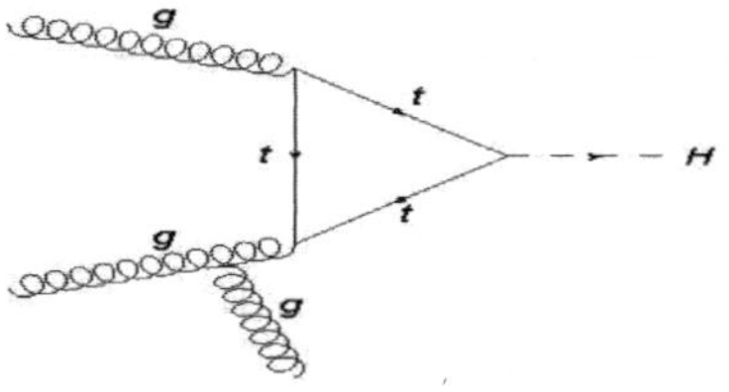

Figure 3: Feynman diagram for partonic process $gg \longrightarrow gH$ [5]

2. Higgs transvers momentum data obtained form the LHC

The transvers momentum data for Higgs production obtained from the LHC for the center of mass energy of 8TeV is shown in figure 4, 5 and 6. The shaded regions depict what theory predicts and the dots represent the actual data obtained. Figure 4 and 5 was obtained from the ATLAS experiment and figure 6 was obtained from the CMS experiment.

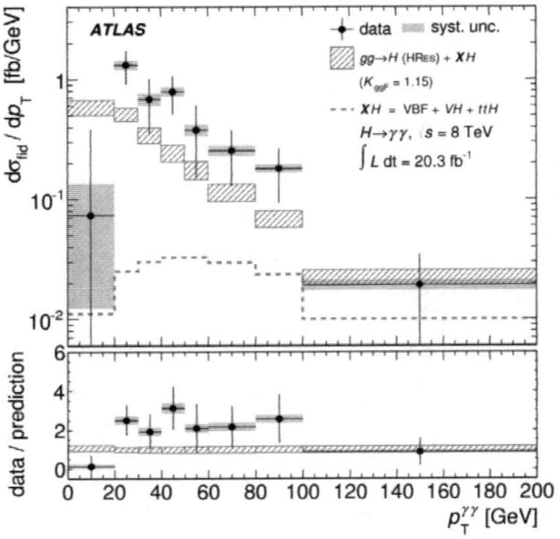

Figure 4: $H \rightarrow \gamma\gamma$ **[6]**

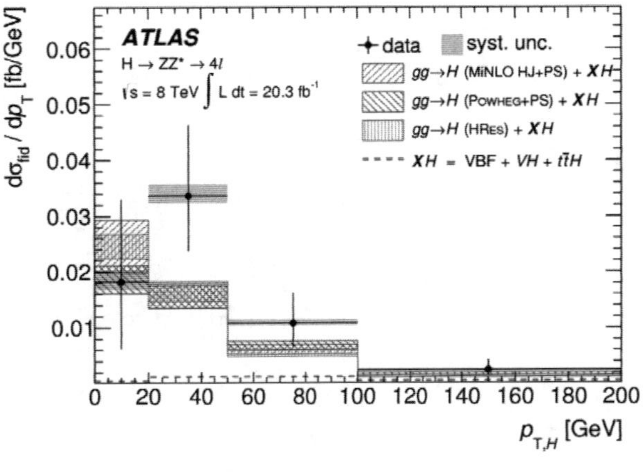

Figure 4: $H \rightarrow ZZ^{*} \rightarrow 4l$ **[7]**

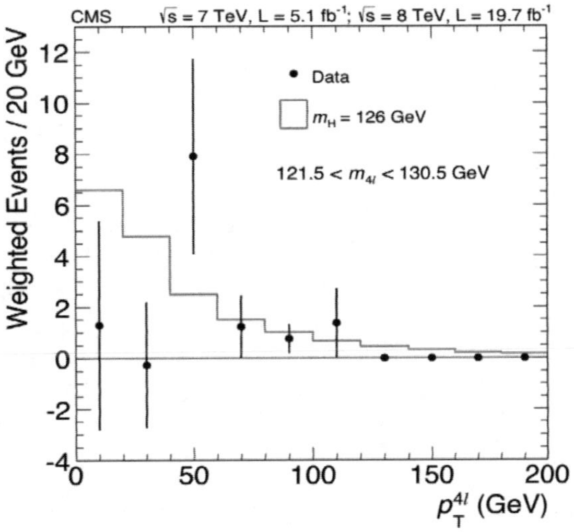

Figure 6: $H \rightarrow ZZ^{*} \rightarrow 4l$ **[8]**

The data collected at ATLAS as well as at CMS show a structure that does not correlate with theory from around 40 to 100 GeV. One can see the data has got higher cross-sections than predicted over this range. This discrepancy could be the result of a statistical fluctuation (i.e. we got lucky) and it doesn't truly represent what happens, an error in the model (i.e. QCD is not properly understood for this process) or it could indicate new physics. The same structure is however observed at both ATLAS and CMS and since they use different physical mechanisms for detection, this makes a statistical fluctuation less likely. Further more the theory so well predicts Z boson production at ATLAS as the theory and the data correlate quite well particularly for the region where we see discrepancies for the Higgs. Which compels us to believe that our model for Higgs production could not be as far off as it is. Thus this structure may indicate new physics.

Journal of Physics: Conference Series **645** (2015) 012010 doi:10.1088/1742-6596/645/1/012010

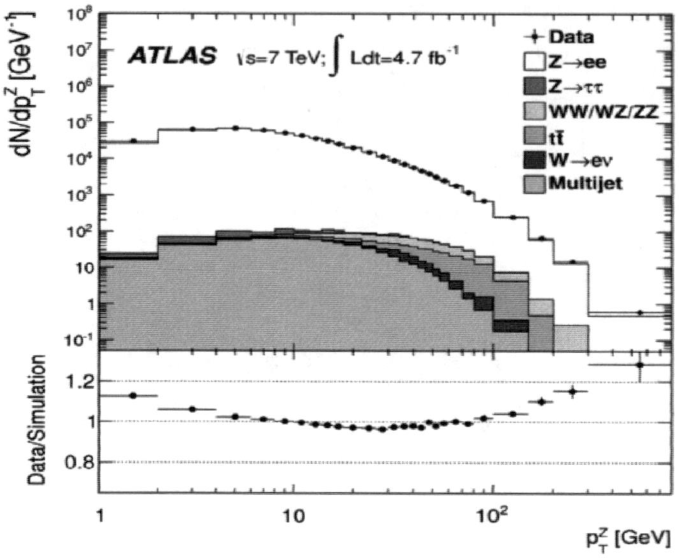

Figure 7: P_T for Z boson production at ATLAS [9]

3. Investigation into the transvers momentum of Higgs production at the LHC

In order to investigate the P_T data for Higgs production and how it will change when the LHC runs at the higher centre of mass energy of 13TeV I ran simulations using MadGraph [10] for Higgs production at the LHC for different centre of mass energies from 8TeV to 14TeV in 1TeV steps. I did this for proton collisions resulting in a Higgs and a jet ($pp \rightarrow h\, j$) as well as two jets ($p\, p \rightarrow h\, j\, j$). I then compared the P_T cross-sectional data, from these simulations, for each energy step to each other. I plotted the P_T cross-sectional data for the respective energies. I also plotted the ratios of this data (figure 8,9), the integrated cross-sectional data and the ratios of the integrated cross-sectional data (figure 10,11). What these plots show is that the cross-section is expected to increase as the energy increase and that the increase will be faster at higher P_T values. These plots are useful in that they give predictions that can be compared to actual data and in this case highlight areas of disagreement with theory and data.

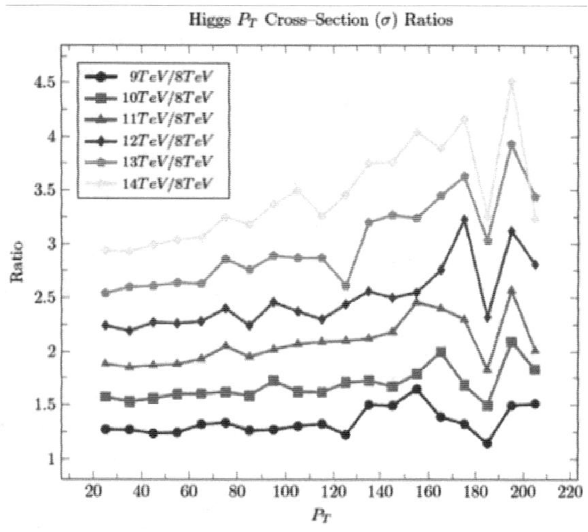

Figure 8: Ratios of P_T cross-sectional data for $pp \rightarrow h\, j$

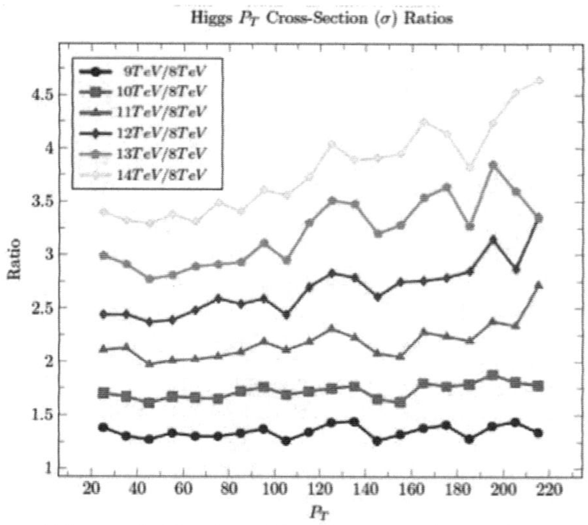

Figure 9: Ratios of P_T cross-sectional data for $pp \rightarrow h\, j\, j$

Journal of Physics: Conference Series **645** (2015) 012010 doi:10.1088/1742-6596/645/1/012010

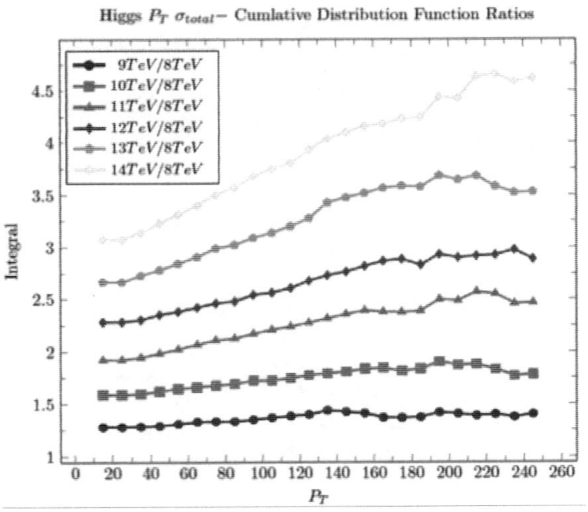

Figure 10: Ratios of P_T integrated cross-sectional data for $pp \rightarrow h\,j$

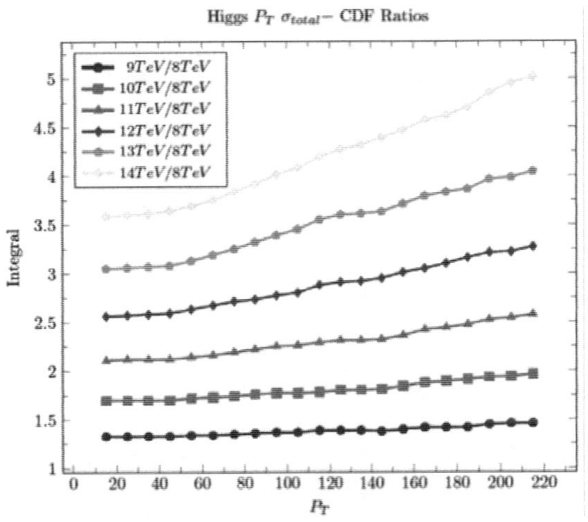

Figure 11: Ratios of P_T integrated cross-sectional data for $pp \rightarrow h\,j$

As discussed above there may be a structure that is not predicted by the SM presenting itself and so it would be useful to investigate models that resolve this discrepancy and investigate how this would influence P_T cross-sectional data for Higgs production and how it may change as the energy changes. A heavy -scalar boson hypothesis mentioned bellow offers a reasonable rationalisation.

3.1. Heavy scalar boson hypothesis

In a few beyond standard model models (BSM) a heavy -scalar boson (A) is predicted which would decay into a SM like Higgs plus another particle. The emission of this other particle would thus give the Higgs more P_T than predicted by the SM and so explain the different structure seen in the P_T data. In order to roughly approximate how this might change the P_T cross-sectional data simulations were done at 8TeV and 13TeV for the process in which protons collide and produce a Higgs ($p\,p \rightarrow h$) where the Higgs is now thought of as an A. Then by varying the mass of the Higgs (A) the ratios of the total cross-sections at 13TeV to 8TeV as a function of the mass of the A was obtained (figure 12). What this plot indicates is that as the mass of A increases the ratio of the total cross-section at 13TeV to 8TeV increases meaning that if there is an A being produced that is decaying into a SM like Higgs and something else, the expectation would be that the amount of Higgs seen at 13TeV would be greater than predicted by the SM.

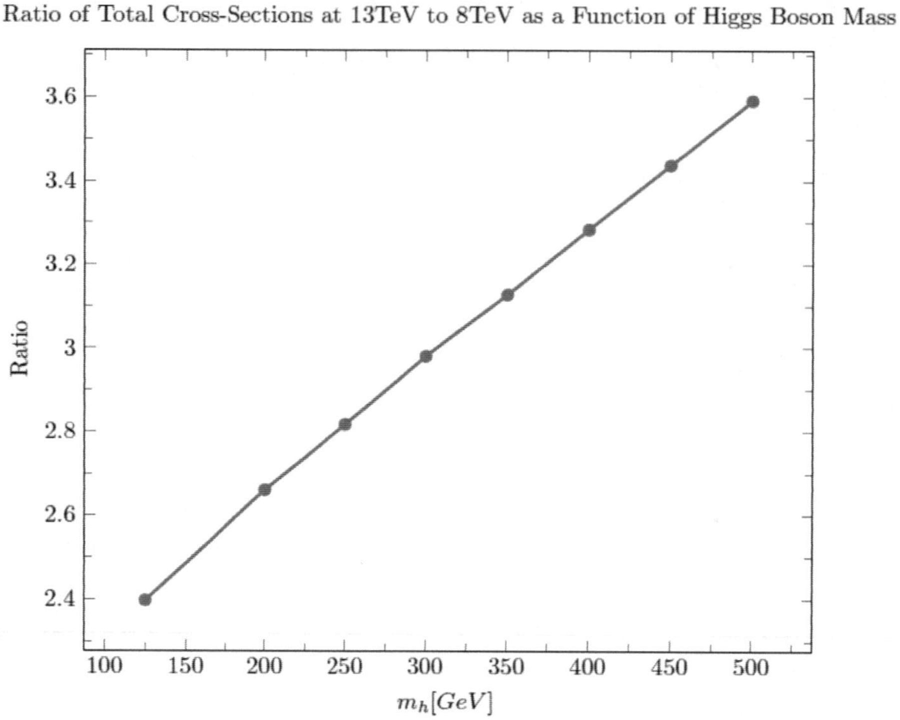

Figure 12: ratio of the total cross-sections at 13TeV to 8TeV as a function of the mass of the heavy scalar boson

References

[1] CERN Bulletin, (2012). 'GO ON A PARTICLE QUEST AT THE FIRST CERN WEBFEST. INTERNET'. http://cds.cern.ch/journal/CERNBulletin/2012/35/News%20Articles/1473657. Accessed 9 September 2014

[2] Robert O, 2005 *The Theory of Almost Everything* New York Pi Press 336 pages

[3] CERN, (2014). 'The Large Hadron Collider'. INTERNET. http://home.web.cern.ch/topics/large-hadron-collider. Accessed 21 Sep. 14

[4] Bentvelsen, S., Laenen, E. & Motylinski, P., (2005). 'Higgs production through gluon fusion at leading order'. *NIKHEF.* 007

[5] Binosi, D. & Theuß, L., (2004). 'JaxoDraw: A graphical user interface for drawing Feynman diagrams', *Comp. Phys. Comm.* 161 (1-2). p.76-86

[6] ATLAS Collaboration (Georges Aad (Marseille, CPPM) *et al.*) 2014 *J. High Energy Phy.* JHEP09(2014)112

[7] ATLAS Collaboration (Georges Aad (Marseille, CPPM) et al.) 2014 *Phys. Lett.* B **738** 234-253

[8] CMS Collaboration (Serguei Chatrchyan (Yerevan Phys. Inst.) et al.) *Phys. Rev.* D **89** 092007

[9] ATLAS Collaboration (Georges Aad (Marseille, CPPM) *et al.*) 2014 *J. High Energy Phy.* JHEP09(2014)145

[10] J. Alwall et al 2014 *J. High Energy Phy.* JHEP07(2014)079

Acknowledgments

I would like to thank Professor Mellado for his help and guidance in this project.

Results of the search for an A boson decaying to Zh, with an $\ell\ell\tau\tau$ final state, in pp collisions at 8 TeV centre of mass energy recorded with the ATLAS experiment

Guillermo Nicolas Hamity
School of Physics, University of the Witwatersrand, Johannesburg 2050, South Africa
E-mail: guillermo.nicolas.hamity@cern.ch

Abstract. The neutral CP-dd boson A is predicted by many models with an extended Higgs sector. Searching for the A boson in the sensitive Zh decay, where h is assumed to be the LHC discovered Higgs boson, within the mass range of $220 - 1000$ GeV, offers a gateway to find physics beyond the Standard Model. A search for a gluon-fusion-produced A in the decay to Zh, with a final state of two light leptons and two tau leptons, is conducted with 20.3 fb^{-1}of proton-proton collision data at 8 TeV center of mass energy. The data driven background estimations, background reduction techniques and systematic uncertainty calculations are presented. Upper limits on the cross section times branching ratio of the A boson decaying to $\ell\ell\tau\tau$ are set for various 2-Higgs-Doublet-Model (2HDM) scenarios. Where no excess is observed, exclusion limits are set on ranges of the 2HDM phase-space.

1. Introduction

The Standard Model (SM) is the centrepiece describing the underlying principles of fundamental particle interactions at high energies. The model combines the fundamental electromagnetic, weak and strong forces, and is able to predict known particle interactions with great precision. The SM has withstood the test of multiple experimental observations, with it's *tour de force* being the recent discovery of the Higgs boson which is predicted in the SM by the ATLAS and CMS experiments at the LHC in Geneva [1, 2]. The discovered Higgs boson was found with a mass around 125 GeV and is thus far consistent with the SM zero-spin and positive-parity hypothesis [3, 4]. It therefore comes as no surprise that present ambiguities with the SM are not phenomenological, but stem rather from the it's 'collage-type' derivation [5]. The repetition of fermionic families and the un-quantification of quark charges suggest that the SM may be a low energy state of some underlying theory. More troublesome is that the SM gives no cause for the imbalance of matter over anti-matter, and excludes the gravitational force and dark matter. Models that describe Beyond the Standard Model (BSM) physics are generally aimed at answering some of these questions. In this light, searching for inconsistencies in the SM is crucial to searching for BSM phenomena.

One of the most basic extensions to the SM is by including an additional Higgs doublet in it's derivation [6]. The SM Higgs sector is only comprised of a simple scalar structure. When considering that the fermionic structure has multiple families and mixings, this assumption

Journal of Physics: Conference Series **645** (2015) 012011 doi:10.1088/1742-6596/645/1/012011

need no be the case. The extended models with two Higgs doublets are intuitively referred to as 2-Higgs-Doublet-Models (2HDMs). They include the Minimal Supersymmetric Standard Model (MSSM), an important candidate of Supersymmetry (SUSY), as well as axion [7] and baryogenesis [8] models. After including an additional Higgs doublet, as well as a few intuitive assumptions [6], 2HDMs give rise to five 'Higgs' bosons. There are two CP-even bosons, the light h and heavy H, distinguishable only by their mass, where $m_h < m_H$. Then there are two charged Higgs bosons, denoted as $H^{\pm}$, and a single CP-odd pseudoscalar, the A boson. One should note that the h and H bosons are candidates for the discovered boson, due to their zero-charge and even-parity nature. This allows for 2HDM searches where the mass of one of the CP-even bosons is that of the discovered Higgs. Generally in 2HDM searches one assumes that $m_h = 125$ GeV, although it is still feasible that the light Higgs lives somewhere in the $m_h < 125$ GeV range. Another common assumption made is that the remaining Higgs bosons have approximately equal masses $m_A \approx m_{H^{\pm}} \approx m_H$ [9]. With these assumptions the 2HDM phasespace is reduced considerably to three free parameters: the mass of one of the heavier Higgs bosons ($e.g\ m_A$), the mixing angle of the two CP-even Higgs bosons, α, and the ratio of the vacuum expectation values of the two Higgs doublets, β. The last two parameters are generally represented as $\tan\beta$ and $\cos(\beta-\alpha)$ for convenience. In direct searches for the 2HDM bosons, all thee parameters are allowed to vary. Furthermore the predicted couplings of the bosons to the vector bosons and fermions fluctuate in terms of $\tan\beta$ and $\cos(\beta-\alpha)$. An additional freedom in terms of the way that each Higgs doublet couples to vector bosons and fermions gives rise to four different 2HDM types. The ratio of the predicted coupling scale factors for the different 2HDM scenarios of the h boson to the SM Higgs boson couplings is given in table 1. Coupling ratios to the vector bosons (κ_V), up-type quarks (κ_u), down-type quark (κ_d), and leptons (κ_ℓ) are shown to vary as functions of β and α.

Table 1. Ratio of predicted coupling scale factors to SM particles of the four 2HDM h bosons to the predicted SM Higgs couplings [13].

Coupling scale factor	Type-I	Type-II	Type-III	Type-IV
κ_V	$\sin(\beta-\alpha)$	$\sin(\beta-\alpha)$	$\sin(\beta-\alpha)$	$\sin(\beta-\alpha)$
κ_u	$\cos(\beta)/\sin(\alpha)$	$\cos(\beta)/\sin(\alpha)$	$\cos(\beta)/\sin(\alpha)$	$\cos(\beta)/\sin(\alpha)$
κ_d	$\cos(\beta)/\sin(\alpha)$	$-\sin(\alpha)/\cos(\beta)$	$\cos(\beta)/\sin(\alpha)$	$-\sin(\alpha)/\cos(\beta)$
κ_ℓ	$\cos(\beta)/\sin(\alpha)$	$-\sin(\alpha)/\cos(\beta)$	$-\sin(\alpha)/\cos(\beta)$	$-\sin(\alpha)/\cos(\beta)$

Direct searches for 2HDM boson set upper limits on the cross section times decay branching ratio of a specific final state channel. In this way we are able to search for the existence of the 2HDM bosons and subsequent traces of BSM physics. As a bi-product of these studies one is able to exclude the existence of 2HDMs from certain areas of the m_A, $\tan\beta$, and $\cos(\beta-\alpha)$ phasespace if no sufficient excess above the upper limits is observed. Various searches have already been conducted, including the CMS searches for $H \to hh$ and $A \to Zh$ between $260-360$ GeV [10], and an independent search for $A \to Zh \to \ell\ell b\bar{b}$ [11]; and the $A \to \tau\tau$ search with ATLAS [12]. No excess was measured in the studies and large sections of the 2HDM phasespace were excluded. Indirect exclusion limits have also been placed on 2HDMs in the ($\tan\beta$, $\cos(\beta-\alpha)$) plane by interpreting the observed couplings of the discovered Higgs boson to other particles [13]. The ($\tan\beta$, $\cos(\beta-\alpha)$) phasespace is scanned for points where the measured Higgs couplings disagree with the 2HDM predictions, at 95% CL or greater, so as to exclude regions in ($\tan\beta$, $\cos(\beta-\alpha)$). There are however areas of phasepace for 2HDMs which have not been probed due to the low branching ratios of the various topologies.

The $A \rightarrow Zh$ channel with the h decaying to $\tau\tau$ and Z to $\ell\ell$ is a new search that promises good sensitivity. In particular this channel probes areas of the Type-II and Type-IV 2HDMs which have not been accessible in previous searches. In this topology three orthogonal final states are available, depending on the either leptonic or hadronic decay of each tau lepton. They are $\tau_{lep}\tau_{lep}$, $\tau_{lep}\tau_{had}$ and $\tau_{had}\tau_{had}$. The production of the pseudoscalar A boson is predominantly driven by gluon-fusion. Contributions from b-associated production only become relevant for Type-II and Type-IV 2HDMs at high values of $\tan\beta$, dominating over gluon-fusion at around $\tan\beta = 10$ [9]. In most cases the ratio of b-associated A boson production to that of gluon-fusion is restricted to $< 3\%$. Two leading order Feynman diagrams of the gluon-fusion and b-associated A productions are shown in figure 1.

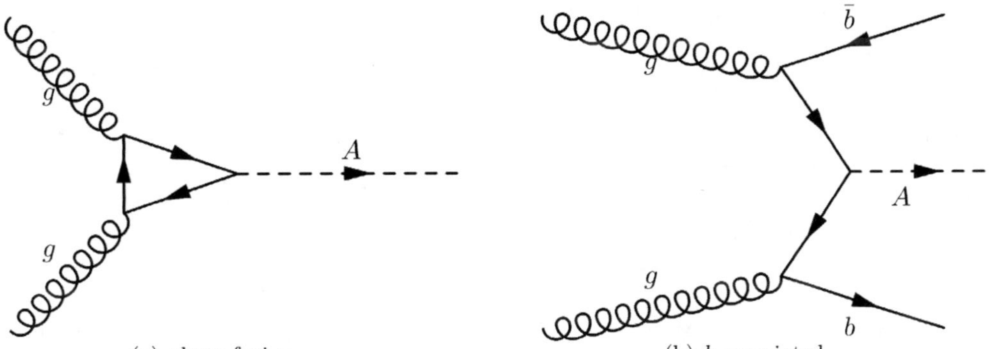

(a) gluon-fusion (b) b-associated

Figure 1. Feynman diagrams for (a) gluon-fusion and (b) b-associated A boson production.

Generally speaking, $A \rightarrow Zh$ is the dominant decay mode of A in any 2HDM when the mass of the A is less than two times the mass of the top-quark ($m_h + m_Z < m_A < 2m_t$). This makes the $A \rightarrow Zh$ channel a very promising search. At masses of $m_A > 2m_t$ the branching ratio BR($A \rightarrow Zh$) drops drastically due to the exposed $t\bar{t}$ kinematic region. In this paper the results from the search of the gluon-produced A boson decaying to $Zh \rightarrow \ell\ell\tau_{had}\tau_{had}$ are presented in the range of $220 < m_A < 1000$ GeV. The results are combined with the $\tau_{lep}\tau_{lep}$ and $\tau_{lep}\tau_{had}$ searches and finally interpreted in the various 2HDM scenarios in conjunction with a complementary $A \rightarrow Zh \rightarrow ffb\bar{b}$ analysis, where $f = \ell$, or ν.

2. The ATLAS detector

The LHC is currently the largest man-made particle accelerator in the world, capable of proton-proton (pp), and heavy ion (lead-proton (Pb-p) and lead-lead (Pb-Pb)) collisions. It has a circumference of 27 km, a design center-of-mass energy of 14 TeV, and peak instantaneous luminosity of 10^{34} cm^{-2}s^{-1} for pp collisions. The ATLAS detector is one of the two general-purpose detectors at the LHC [14]. It was designed to accommodate a wide spectrum of different physics signatures and explore processes stemming from the TeV mass scale, where groundbreaking discoveries are expected. The ATLAS detector is a multi-purpose particle detector with approximately forward-backward symmetric cylindrical geometry. ATLAS uses a right handed coordinate system with the origin at the interaction point. The z-axis is along the beam pipe, x-axis points towards the center of the LHC ring and the y-axis points up. Polar coordinates (r, ϕ) are used in the transverse plane, ϕ being the azimuthal angle around the beam pipe. The pseudorapidity η is defined as $\eta = -\ln(\tan(\theta/2))$ where θ is the angle between the particle momentum 3-vector and the beam axis. The inner detector (ID) covers $|\eta| < 2.5$ and consists of silicon pixels, silicon micro-strips and a transition radiation tracker. The ID is surrounded by a superconducting solenoid providing a 2 T magnetic field. Surrounding the

ID are two calorimeters, the inner electromagnetic (EM) and the outer hadron calorimeters. The EM calorimeter measures the energy and the position of EM showers within $|\eta| < 3.2$ as well as hadronic showers in the end-cap ($1.5 < |\eta| < 3.2$) and forward ($3.1 < |\eta| < 4.9$) regions. The hadronic calorimeter measures hadronic showers in the central region ($|\eta| < 1.7$). The muon spectrometer surrounds the calorimeters and consists of three subsystems: eight superconducting air-core toroid magnets, a system of tracking chambers and a fast tracking chamber for triggering. A three-level trigger system selects events to be recorded for offline analysis.

3. The $A \to Zh \to \ell\ell\tau_{had}\tau_{had}$ analysis

A search for $A \to Zh(\tau\tau)$ is conducted with the 2012 ATLAS data at 8 TeV center of mass energy, where the light h boson is assumed to have a mass of $m_h = 125$ GeV. Signal Monte Carlo (MC) samples are produced with Pythia8 [15] for A masses at fixed working points: $m_A = 220, 240, 260, 300, 340, 350, 400, 500, 800, 1000$ GeV. As mentioned in the introduction, each tau in the analysis can decay leptonically (35%) or hadronically (65%). Since tau leptons are massive particles, they decay quickly in the detector and are never observed directly in ATLAS, but can only be reconstructed from their decay products. This makes differentiating light-leptons originating from leptonic-tau-decays to light-leptons originating from other processes virtually impossible. As such, tau leptons in the ATLAS detector are considered to be hadronically decaying taus. Tau leptons decay hadronically into either 1− or 3−charged pions, with a ν_τ and possibly a neutral pion. Since the BR($A \to Zh$) is already relitavely small, and since the SM branching ratio BR($h \to \tau\tau$) is around 1000 times larger that BR($h \to \mu\mu$), searches in the $\tau_{lep}\tau_{lep}$ and $\tau_{lep}\tau_{had}$ channels can be conducted in $\tau\tau$ without worrying about contamination from $h \to \ell\ell$ ($\ell \neq \tau$). In all three analysis stradegies, the reconstructed A boson mass, m_A, is considered as the parameter of interest.

The fully hadronic final state is a particularly challenging analysis since it contains a large amount of missing transverse energy, E_T^{miss}, from the two neutrinos. In the case of $\tau_{had}\tau_{had}$, full reconstruction of the event topology requires solving a system of equations with 6 unknowns. This system is under-constrained and an exact solution is not possible. Therefore, the full event reconstruction, as well as the full reconstruction of the $m_{\tau\tau}$ mass, are impossible. The Missing Mass Calculator (MMC) method is therefore used to reconstruct $m_{\tau\tau}$ [16]. This method scans for the most likely $m_{\tau\tau}$ solution given the known conditions of the event, *i.e.* the visible taus and E_T^{miss} kinematics. The MMC is compared to alternate $m_{\tau\tau}$ reconstruction methods: the transverse mass and collinear approximation (CLA) methods [16], performing at a better efficiency and resolution than the CLA method.

The event selection in the $\tau_{had}\tau_{had}$ channel focuses on selecting a clean signal with exactly two leptons (e, or μ) and two visible tau jets. Events are triggered by single lepton triggers. The events are required to have exactly two same flavour opposite charge loose leptons with a combined invariant mass of $80 < m_{\ell\ell} < 100$ GeV, which corresponds to $m_Z = 90$ GeV. Track and calorimeter track isolation requirements are also imposed on the leptons. The event must also have exactly two opposite charge loose tau leptons, which need to pass electron and muon vetos, and have a MMC combined invariant mass of $75 < m_{tautau}^{\text{MMC}} < 175$ GeV, corresponding to $m_h = 125$ GeV. The large window here compensates for the resolution of the MMC algorithm. Two additional selection conditions are applied after an optimization study performed on different kinematic variables. The first of these is that the tau lepton with highest p_T should have a transverse energy of $E_T > 35$ GeV. Finally, events where the reconstructed di-lepton p_T is smaller than $0.64 \times m_A - 131$ GeV are discarded. This selection is applied to all events with a reconstructed A mass less than 400 GeV.

After the full event selection, the $\tau_{had}\tau_{had}$ analysis has a few surviving background processes, the most prominent of which is the Z+jets background. This background contains jets which are mis-identified as tau leptons. A small contribution from WZ di-boson processes also contributes to the fake background. Other important backgrounds include di-boson ZZ and SM ZH production, which are comprised of true tau signatures, and are estimated from MC. All backgrounds caused by mis-identified jets are estimated via a data-driven template method prediction. This method is performed by constructing three orthogonal control regions. The first region is defined identically to the full signal selection, except that the leading tau is forced to fail the loose tau selection. The second region requires that the tau leptons have the same charge. The third region combines both the inverted loose tau and same tau charge requirements. When looking at data in these regions, the control regions are completely populated by fake tau jets. Data distributions in the control regions are extrapolated into the signal region and used to estimate the mis-identified tau background. The distribution is normalized using m_A sidebands.

Systematic uncertainties due to the fake background normalization are considered by taking into account the fluctuations in normalization values when choosing alternate control regions. This uncertainty constitutes the largest uncertainty, approximately 25%. Uncertainties due to detector calibration effects, and theoretical uncertainties are also included. The m_A distribution of the $\tau_{had}\tau_{had}$ channel after full event selection is given in figure 2, showing the predicted backgrounds, the 340 GeV mass signal signal with a cross section times branching ratio of 50 fb, and the observed data.

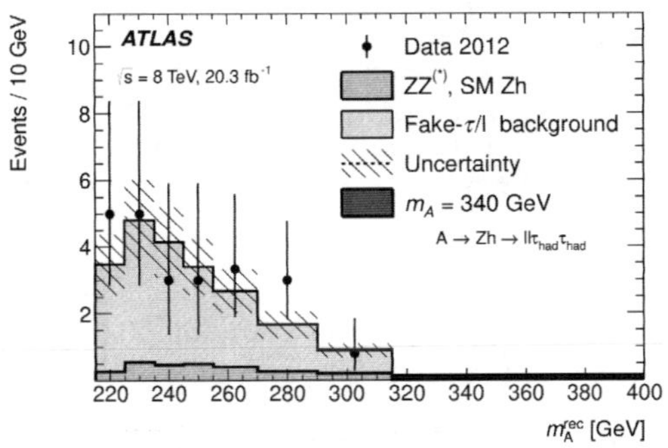

Figure 2. The reconstructed A boson mass, m_A of the $\tau_{had}\tau_{had}$ channel after full event selection [17].

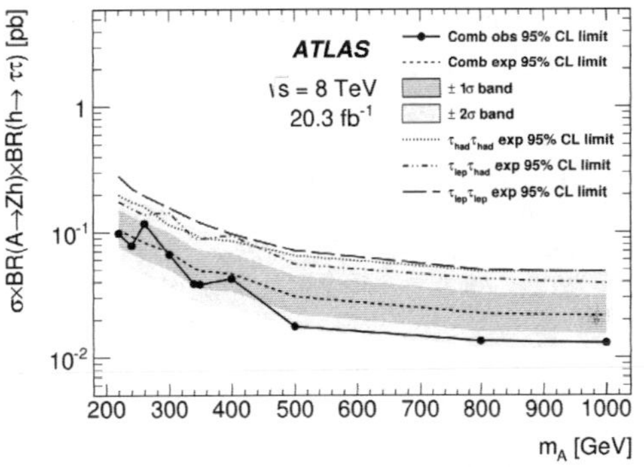

Figure 3. Expected and observed 95% CL upper limits of the gluon-fusion cross section times branching ratio of the combined $A \to Zh \to \ell\ell\tau\tau$ [17].

4. Results and Conclusions

A binned likelihood ratio test, along with an asymptotic approximation, is used in order to plot the 95% CL upper limits on the cross section times branching ratio of $\ell\ell\tau\tau$. Figure 3 shows the expected and observed upper limits of gluon-fusion $\sigma \times \mathrm{BR}(A \to Zh) \times \mathrm{BR}(h \to \tau\tau)$ for the combined $A \to Zh \to \ell\ell\tau\tau$ analysis. No assumptions are made on the branching ratio of $h \to \tau\tau$ in order to maintain the result model independent. The observed combined upper limit is within 2σ of the expected limit. The expected 95% CL for the $\tau_{lep}\tau_{lep}$, $\tau_{lep}\tau_{had}$ and $\tau_{had}\tau_{had}$ are shown in dashed lines. The drop in the limits at higher values are from the lack of data observed in the highest m_A bins. The 95% CL upper limits on the gluon-fusion $\sigma \times \mathrm{BR}(A \to Zh) \times \mathrm{BR}(h \to \ell\ell\tau\tau)$ range from 0.098–0.013 pb in a range of $220 \leq m_A \leq 1000$ GeV. No excess is observed outside the 2σ. The low observed limits at the higher mass points is caused by a lack of events in the final 315-2000 GeV m_A^{rec} bin.

The final limits are combined with limits from a complementary $A \rightarrow Zh \rightarrow ff b\bar{b}$ analysis, where $f = \ell, \nu$ [17]. They are interpreted in the context of the four 2HDM scenarios in order to exclude sections of the phasespace. The final paper includes the first published search for an A boson including the $\ell\ell\tau\tau$ channel.

References

[1] ATLAS Collaboration, 2013 Measurements of Higgs boson production and couplings in diboson final states with the ATLAS detector at the LHC *Phys.Lett. B* **726** 88
[2] CMS Collaboration, 2013 Combination of standard model Higgs boson searches and measurements *CMS-PAS-HIG-13-005*
[3] ATLAS Collaboration, 2013 Evidence for the spin-0 nature of the Higgs boson using ATLAS data *Phys.Lett. B* **726** 120
[4] CMS Collaboration, 2013 Study of the Mass and Spin-Parity of the Higgs Boson Candidate Via Its Decays to Z Boson Pairs *Phys.Rev.Lett.* **110** 081803
[5] Langacker P, 1981 Grand unified theories and proton decays *Phys.Rev.Lett.* **72** 185
[6] Branco G C, *et al.*, 2012 Theory and phenomenology of two-Higgs-doublet models *Phys.Rept.* **516** 1
[7] Kim J E, 1987 Light Pseudoscalars, Particle Physics and Cosmology *Phys.Rept.* **150** 1
[8] Joyce M, Prokopec T and Turok N, 1994 Nonlocal electroweak baryogenesis. Part 2: The Classical regime *Phys.Rev.* **D53** 2958
[9] ATLAS Collaboration, 2013 Beyond-the-Standard-Model Higgs boson searches at a High-Luminosity LHC with ATLAS *ATLAS-PUB-2013-016*
[10] CMS Collaboration, 2013 Search for extended Higgs sectors in the H to hh and A to Zh channels in $\sqrt{(s)} = 8$ TeV pp collisions with multileptons and photons final states *CMS-PAS-HIG-13-025*
[11] CMS Collaboration, 2014 Search for a pseudoscalar boson A decaying into a Z and an h boson in the llbb final state *CMS-PAS-HIG-14-011*
[12] ATLAS Collaboration, 2014 Search for neutral Higgs bosons of the minimal supersymmetric standard model in pp collisions at $\sqrt{s} = 8$ TeV with the ATLAS detector *J. High Energy Phys.* JHEP11(2014)056
[13] ATLAS Collaboration, 2014 Constraints on New Phenomena via Higgs Coupling Measurements with the ATLAS Detector *ATLAS-CONF-2014-010*
[14] CERN, 1999 ATLAS: Detector and physics performance technical design report. Volume 1 *CERN-LHCC-99-14*
[15] Sjostrand T, Mrenna S and Skands P, 2006 PYTHIA 6.4 Physics and Manual *JHEP* **0605** 026
[16] Elagin A, Murat P, Pranko A and Safonov A, 2011 A New Mass Reconstruction Technique for Resonances Decaying to di-tau *Nucl.Instrum.Meth.* **A654** 481
[17] ATLAS Collaboration, 2015 Search for a CP-odd Higgs boson decaying to Zh in pp collisions at $\sqrt{s} = 8$ TeV with the ATLAS detector *CERN-PH-EP-2015-013*

Journal of Physics: Conference Series **645** (2015) 012012 doi:10.1088/1742-6596/645/1/012012

Missing p_{T} Reconstruction at ATLAS

Claire A. Lee
University of Johannesburg, South Africa
E-mail: `claire.lee@cern.ch`

Abstract. The missing transverse momentum ($E_{\mathrm{T}}^{\mathrm{miss}}$) in particle collider experiments is defined as the momentum imbalance in the plane transverse to the beam axis: the resultant of the negative vectorial sum of the momenta of all the particles that are involved in the *pp* collision of interest. A precise measurement of the $E_{\mathrm{T}}^{\mathrm{miss}}$ is essential for many physics studies at the LHC, such as Higgs boson searches and measurements, as well as searches beyond the Standard Model. The $E_{\mathrm{T}}^{\mathrm{miss}}$ measurement is constructed from the reconstructed and calibrated energy deposits inside the calorimeters, a method that has historically served experiments well, but one which is sensitive to fluctuations from noise and, in particular, additional unrelated collisions within the same event - an effect that is becoming more critical with the increasing luminosity of the LHC. A complementary method for measuring the missing transverse momentum is presented, in which track momenta are used in place of the calorimeter energy measurements, allowing the calculation to be made from particles originating solely from the collision vertex of interest. The reconstruction of this track-based missing transverse momentum, $p_{\mathrm{T}}^{\mathrm{miss}}$, and its performance in W and Z boson events, is described here.

1. Introduction

Missing transverse momentum ($E_{\mathrm{T}}^{\mathrm{miss}}$) is an essential part of many physics analyses at the LHC, including Higgs boson measurements and searches for beyond the Standard Model (BSM) signatures such as supersymmetric or hidden sector particles. After the *pp* scattering, most Standard Model (SM) particles produced leave traces inside the detector, however, SM neutrinos and some theorised BSM particles escape the detector without producing a signal. Since the momenta of the colliding protons in the transverse direction is essentially zero (in comparison to their boost along the beam direction), the presence of these particles can be inferred by a resultant momentum imbalance in the transverse plane. Ideally the $E_{\mathrm{T}}^{\mathrm{miss}}$ should be calculated using the final state particles from the hard *pp* interaction only (defined in ATLAS as the vertex with the highest scalar $\sum(p_T^{track})^2$), summing the momentum contributions from both charged and neutral particles over the full 4π solid angle. In practice, however, the detector phase space, reconstruction efficiencies, and the presence of additional "pileup" interactions from other *pp* scattering vertices in the same bunch crossing event, all affect the measurement.

2. The ATLAS Detector

The ATLAS detector [1] is a multipurpose system of particle detectors with nearly 4π solid angle coverage. It is composed of 3 core systems arranged in a barrel-plus-endcaps format. The inner detector (ID), covering the pseudorapidity range $|\eta| < 2.5$[1], consists of a 3-layered,

[1] In ATLAS, the positive x-axis is defined as pointing from the interaction point to the center of the LHC ring, the positive y-axis is defined as pointing upwards, and the positive z-axis corresponds to protons running

high granularity silicon pixel detector, followed by a silicon microstrip detector (SCT) with 8 layers, providing four two-dimensional measurement points per track. Additionally, a transition radiation tracker (TRT) adds extra tracking and electron identification ability within the $|\eta| < 2$ range. Surrounding the ID, a high granularity lead/liquid argon (LAr) sampling electromagnetic calorimeter covers the region $|\eta| < 3.2$ while a steel/scintillator-tile calorimeter provides hadronic coverage within $|\eta| < 1.7$. A Copper-LAr hadronic calorimeter is used in the end-cap region $1.5 < |\eta| < 3.2$. In the forward region, $3.1 < |\eta| < 4.9$, a copper-LAr electromagnetic calorimeter and a copper/tungsten-LAr hadronic calorimeter for a full energy measurement. Finally, the muon spectrometer (MS) surrounds the calorimeters, consisting of high-precision monitored drift tubes and cathode strip chambers for tracking in the barrel and endcaps respectively, providing accurate muon momentum measurements out to $|\eta| < 2.7$, as well as resistive plate chambers (barrel) and thin gap chambers (endcaps) for muon triggering in the region $|\eta| < 2.4$.

3. Missing Transverse Momentum Reconstruction at ATLAS

3.1. Calorimeter-based reconstruction

The primary detectors used for the $E_{\mathrm{T}}^{\mathrm{miss}}$ measurement in ATLAS are the electromagnetic and hadronic calorimeters, which are designed to provide excellent energy resolution at the LHC scale, and an optimised reconstruction and calibration procedure was developed by the collaboration for use during the LHC Run 1, described in detail in [2]. Calorimeter energy deposits are associated with high p_{T} objects, creating a term for each object type from their negative vectorial sum. Those not associated with high p_{T} objects are also taken into account, grouped into a "soft term" which includes contributions from both low p_{T} (<20 GeV) jets, and unassociated topoclusters. The $E_{\mathrm{T}}^{\mathrm{miss}}$ is then calculated as the vectorial sum of each of the terms. Another often-used quantity for parametrisation and event classification is the total transverse energy of the event, $\Sigma \mathrm{E}_{\mathrm{T}}$, defined by the scalar sum of the same components. The advantage of this method is that it provides a complete measurement of all types of particles involved in the pp interaction. However, energy contributions from pileup vertices can provide sources of "fake" $E_{\mathrm{T}}^{\mathrm{miss}}$ and decrease the resolution significantly. To combat this, the precise tracking and vertexing ability of the ID is taken advantage of.

3.2. Track-based reconstruction

Similar to the calorimeter-based method, individual terms for electrons, jets, and muons are formed, except that here the terms instead use the momenta of the ID tracks associated with the leptons and jets. In place of the calorimeter cluster-based "soft term", only tracks originating from the signal vertex of interest that have not already been included in the lepton or jet terms are used, forming a "track soft term". The nominal $p_{\mathrm{T}}^{\mathrm{miss}}$ is then:

$$\mathbf{p}_{x,y}^{\mathrm{miss,nominal}} = -\left(\sum_{electron\ tracks} \mathbf{p}_{x,y} + \sum_{muon\ tracks} \mathbf{p}_{x,y} + \sum_{jet\ tracks} \mathbf{p}_{x,y} + \sum_{soft\ tracks} \mathbf{p}_{x,y}\right) \tag{1}$$

with

$$\mathbf{p}_{\mathrm{T}}^{\mathrm{miss,nominal}} = \sqrt{\left(\mathbf{p}_{x}^{\mathrm{miss,nominal}}\right)^2 + \left(\mathbf{p}_{y}^{\mathrm{miss,nominal}}\right)^2} \tag{2}$$

with the advantage that it provides a measurement that is almost completely decoupled from the $E_{\mathrm{T}}^{\mathrm{miss}}$, and relates to the signal vertex only. The disadvantage of this measurement, of course, is

anticlockwise. The polar angle θ is measured from the beam axis (z-axis), the azimuthal angle ϕ is measured in the transverse (xy)-plane, and the pseudorapidity is defined by $\eta = -\log(\tan(\theta/2))$.

that it is restricted only to charged particles, as well as by the smaller acceptance of the ID. For certain event topologies the effect is minor, but in events with jets where a significant fraction of the jet momentum comes from neutral particles, or where high p_{T} objects are boosted in the forward directions, missing their momenta leads to a significant decrease in the resolution. A second, object-corrected $p_{\mathrm{T}}^{\mathrm{miss}}$ is therefore defined, where the terms containing the associated tracks for each high p_{T} object are replaced by their reconstructed and fully calibrated momentum counterpart, while retaining the track soft term. In this way, we are able to account for the full p_{T} of jets in both the central and forward regions, electron energies without any radiation losses, and a more accurate high p_{T} muon momentum reconstruction, while maintaining the relation to the signal vertex.

3.3. Track and high p_{T} object selection criteria

All tracks included in the $p_{\mathrm{T}}^{\mathrm{miss}}$ track soft term reconstruction must satisfy the following criteria:

- $p_T > 500$ MeV
- $|\eta| < 2.5$
- At least 1 pixel detector hit
- At least 6 SCT hits
- Transverse impact parameter with respect to the primary vertex $|d_0| < 1.5$ mm
- Longitudinal impact parameters with respect to the primary vertex $|z_0 \times \sin(\theta)| < 1.5$ mm.

The selections ensure that the tracks have enough points for an accurate p_{T} measurement, are associated to the signal vertex, and are within the ID acceptance. Additional selection criteria for specialised cases are also performed, such as removing unwanted tracks around high p_{T} electrons (due to misreconstruction) and badly measured tracks in dense, high p_{T} jets. General tracks with extremely high momenta are also checked against corresponding calorimeter cluster deposits to reduce the number of misreconstructed tracks.

Electrons, muons and jets are selected to complement the method used for the $E_{\mathrm{T}}^{\mathrm{miss}}$. Good quality muons with $p_{\mathrm{T}} > 6$ GeV reconstructed from combined ID and MS tracks, and isolated electrons with $p_{\mathrm{T}} > 10$ GeV are considered for their respective terms. The tracks associated with the selected leptons are obtained using the stored track-to-object information in the data. Jets are reconstructed with the anti-kt algorithm [3] using a distance parameter of R = 0.4, with corrections for pileup at the topocluster level, and subsequent calibrations using the LCW+JES scheme [4, 5]. Jets are selected if their calibrated p_{T} is greater than 25 GeV in the central region ($|\eta| < 2.5$), or 30 GeV in the forward region ($2.5 < |\eta| < 4.4$), and they do not overlap with any of the selected electrons. In addition, a cut on the jet vertex fraction (JVF [6]) of JVF > 0.50 is required for jets with calibrated $p_T < 50$ GeV and $|\eta| < 2.4$, which helps to reduce the selection of pileup jets in the central region. The jet tracks are obtained using the ghost association algorithm [7], and are also required to pass the high quality track requirements described above.

4. $p_{\mathrm{T}}^{\mathrm{miss}}$ Performance in W $\rightarrow \ell\nu$ and Z $\rightarrow \ell\ell$ events

W and Z boson production is relatively abundant at the LHC, and their leptonic decays provide clear experimental signatures with well understood topologies that make them ideal events for measuring the performance of our detector and reconstruction algorithms. The absence of real $E_{\mathrm{T}}^{\mathrm{miss}}$ in Z $\rightarrow \ell\ell$ events allows us to measure the intrinsic resolution and performance of the $p_{\mathrm{T}}^{\mathrm{miss}}$ that results mainly from detector response and object reconstruction efficiencies, whereas in W $\rightarrow \ell\nu$ events the neutrino provides a good metric against which to measure the $p_{\mathrm{T}}^{\mathrm{miss}}$ scale and direction.

Journal of Physics: Conference Series **645** (2015) 012012 doi:10.1088/1742-6596/645/1/012012

4.1. Datasets used and W $\rightarrow \ell\nu$ *and* Z $\rightarrow \ell\ell$ *event selection*

The dataset used for this study was recorded by the ATLAS detector in the 2012 run at a $\sqrt{s}$ = 8 TeV centre of mass energy while all detector subsystems were operating efficiently, with a total integrated luminosity of 20.3 fb^{-1}. Events were required to pass standard detector quality assessment criteria and fire a single or dilepton trigger to be considered. For good vertex reconstruction, the primary vertex is required to have at least three associated reconstructed tracks with $p_T > 400$ MeV each. For Z $\rightarrow \ell\ell$ events, events were selected if they contained exactly two same flavour, opposite charge good quality leptons with $p_{\mathrm{T}} > 25$ GeV, and a combined invariant mass of between 66 and 116 GeV. For W $\rightarrow \ell\nu$ events, exactly one good quality lepton with $p_{\mathrm{T}} > 25$ GeV, and an $E_{\mathrm{T}}^{\mathrm{miss}} > 30$ GeV was required. Furthermore, the invariant mass of the lepton-$E_{\mathrm{T}}^{\mathrm{miss}}$ system was calculated and required to be at least 50 GeV.

Monte Carlo samples of Z $\rightarrow \ell\ell$ and W $\rightarrow \ell\nu$ + n jets ($\ell = e, \mu, \tau$) production were generated using ALPGEN [8] interfaced to Jimmy [9], with up to 6 partons in the final state. Additional background samples of $t\bar{t}$ events were produced with the MC@NLO event generator [10], and WW, WZ, and ZZ diboson samples were produced using HERWIG [11].

4.2. $p_{\mathrm{T}}^{\mathrm{miss}}$ *data vs simulation comparisons*

Data versus Monte Carlo simulation comparisons for the nominal and object-corrected $p_{\mathrm{T}}^{\mathrm{miss}}$ in inclusive Z $\rightarrow ee$ events are shown in figure 1, showing signal and relevant background Monte Carlo distributions ordered by their (weighted) integral so that all background distributions are visible, as well as the data-to-background ratio in the panel below. The error on the ratio points is calculated from the statistical uncertainty of the observed data, and the grey error bands show the total uncertainty from the expected background. A good comparison is obtained between the data and the simulated backgrounds for both the nominal and object-corrected $p_{\mathrm{T}}^{\mathrm{miss}}$, particularly in the peaks, but with some discrepancy in the tails where statistics are low. Some background contribution from QCD processes in these events is also expected, particularly in the low $p_{\mathrm{T}}^{\mathrm{miss}}$ region. These processes are not well simulated and therefore are not included in the background set, giving rise to the discrepancy in the data/background ratio in this region. A peak close to zero is seen as expected for events with no real $E_{\mathrm{T}}^{\mathrm{miss}}$. The broader peak in the nominal distribution is a result of mismeasured electron track p_{T} due to Bremsstrahlung, an effect which is mitigated in the object-corrected case. Also noticeable is the reduced tail in the object-corrected distribution, as the momenta of any jets in the event is fully taken into account. Similar distributions are obtained for the Z $\rightarrow \mu\mu$ channel, though the corresponding nominal distribution does not, of course, have the broader peak.

4.3. $p_{\mathrm{T}}^{\mathrm{miss}}$ *resolution and scale*

The resolution of the $p_{\mathrm{T}}^{\mathrm{miss}}$ is defined as the width of the peak in the x and y component distributions, and is sensitive to both detector and reconstruction efficiencies, as well as external factors like the presence of pileup. The resolution of the nominal and object-corrected $p_{\mathrm{T}}^{\mathrm{miss}}$ is flat as a function of the number of reconstructed primary vertices in the event (N_{pv}), and the values for W $\rightarrow \ell\nu$ and Z $\rightarrow \ell\ell$ events are shown in table 1. The resolution can also be parameterised as a function of the event activity, using:

$$y = k\sqrt{\sum E_T} + c \tag{3}$$

where k and c are constants obtained from a fit of the above function. The resolution for Z $\rightarrow ee$ and W $\rightarrow e\nu$ events as a function of $\sqrt{\sum E_T}$ is shown in figure 2 (a), and the k-values obtained from the fits to the data are also given in table 1 (c is ideally 0 and, in these fits, effectively 0, and so is not shown).

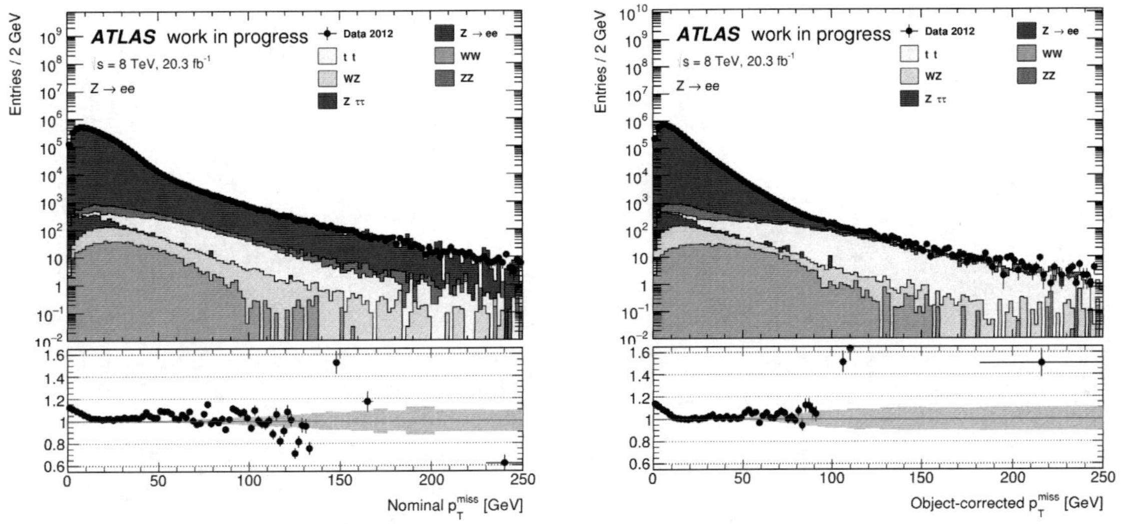

(a) (b)

Figure 1: Data-to-Monte Carlo comparisons of the nominal (a) and object-corrected (b) $p_{\mathrm{T}}^{\mathrm{miss}}$in Z $\rightarrow ee$ events.

Table 1: $p_{\mathrm{T}}^{\mathrm{miss}}$ resolution and $\Sigma\mathrm{E_T}$ fit values

	Z $\rightarrow ee$ Data	Z $\rightarrow ee$ MC	W $\rightarrow e\nu$ MC
Nominal $p_{\mathrm{T}}^{\mathrm{miss}}$ resolution vs N_{pv}[GeV]	10.49	10.33	9.63
Object-corrected $p_{\mathrm{T}}^{\mathrm{miss}}$ resolution vs N_{pv}[GeV]	7.54	7.23	7.15
Nominal $p_{\mathrm{T}}^{\mathrm{miss}}$ fit to ΣE_T (k)	0.28	0.26	0.21
Object-corrected $p_{\mathrm{T}}^{\mathrm{miss}}$ fit to ΣE_T (k)	0.28	0.26	0.24
	Z $\rightarrow \mu\mu$ Data	Z $\rightarrow \mu\mu$ MC	W $\rightarrow \mu\nu$ MC
Nominal $p_{\mathrm{T}}^{\mathrm{miss}}$ resolution vs N_{pv}[GeV]	7.63	7.40	7.12
Object-corrected $p_{\mathrm{T}}^{\mathrm{miss}}$ resolution vs N_{pv}[GeV]	7.51	7.28	7.02
Nominal $p_{\mathrm{T}}^{\mathrm{miss}}$ fit to ΣE_T (k)	0.27	0.26	0.20
Object-corrected $p_{\mathrm{T}}^{\mathrm{miss}}$ fit to ΣE_T (k)	0.29	0.28	0.22

In events with real $E_{\mathrm{T}}^{\mathrm{miss}}$ the scale of the measurement is also important. The linearity of the $p_{\mathrm{T}}^{\mathrm{miss}}$ with respect to the true $E_{\mathrm{T}}^{\mathrm{miss}}$ in an event is defined as the mean of the ratio:

$$\text{Linearity} = \langle \frac{p_{\mathrm{T}}^{\mathrm{miss}} - E_{\mathrm{T}}^{\mathrm{miss,truth}}}{E_{\mathrm{T}}^{\mathrm{miss,truth}}} \rangle \tag{4}$$

which gives an expected value of zero if the $p_{\mathrm{T}}^{\mathrm{miss}}$ is reconstructed at the correct scale. The linearity distributions for the nominal and object-corrected $p_{\mathrm{T}}^{\mathrm{miss}}$ is shown in figure 2 (b) for both W $\rightarrow e\nu$ and W $\rightarrow \mu\nu$ events. A positive bias is visible in the region of very low true $E_{\mathrm{T}}^{\mathrm{miss}}$, owing to the finite resolution of the $p_{\mathrm{T}}^{\mathrm{miss}}$ and the domination of contributions in the soft track term. At larger values of true $E_{\mathrm{T}}^{\mathrm{miss}}$, the bias is within 5% for the object-corrected $p_{\mathrm{T}}^{\mathrm{miss}}$, with a larger negative bias in the nominal $p_{\mathrm{T}}^{\mathrm{miss}}$ as expected due to jets. Differences in the linearity

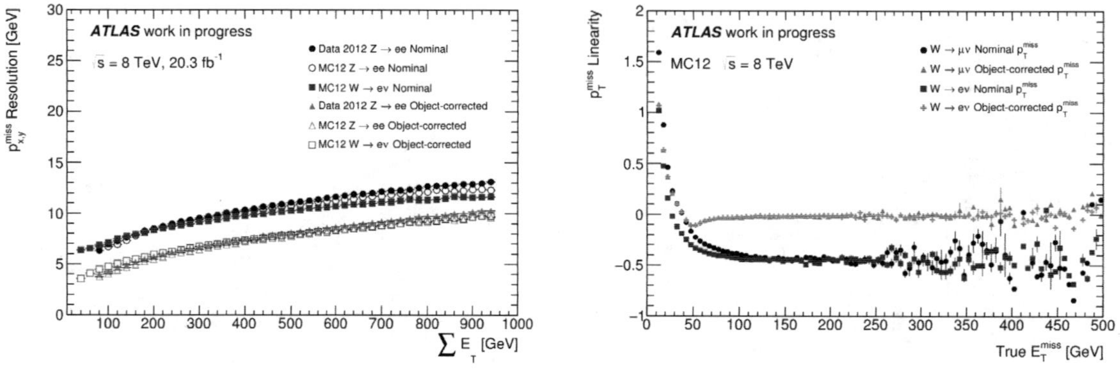

(a) (b)

Figure 2: Resolution of the x and y components of the nominal and object-corrected $p_{\mathrm{T}}^{\mathrm{miss}}$ for $\mathrm{Z} \to ee$ and $\mathrm{W} \to e\nu$ events as a function of the scalar sum of the total transverse energy in the calorimeter. (b) Nominal and object-corrected $p_{\mathrm{T}}^{\mathrm{miss}}$ linearity with respect to the true $E_{\mathrm{T}}^{\mathrm{miss}}$ in $\mathrm{W} \to e\nu$ and $\mathrm{W} \to \mu\nu$ events

for the electron and muon channels can also be seen for the nominal $p_{\mathrm{T}}^{\mathrm{miss}}$, where the electron momentum loss shows again in a worsening of the bias (though the effect at higher values of true $E_{\mathrm{T}}^{\mathrm{miss}}$ is diminished).

5. Conclusions

The measurement of missing energy is an important contribution to a number of interesting physics analyses. The ATLAS calorimeter provides excellent energy resolution but is highly sensitive to pileup. A track-based $p_{\mathrm{T}}^{\mathrm{miss}}$ calculation can be made using the momentum of tracks inside the ATLAS ID, taking advantage of the fact that ID tracks can be selected from a single primary vertex, allowing for much greater pileup stability than the calorimeter-based measurement. The performance of the nominal $p_{\mathrm{T}}^{\mathrm{miss}}$ degrades in events with jets owing to the larger fraction of neutrals and the fiducial coverage, and in events with electrons due to Bremsstrahlung. This is improved by using the fully reconstructed object term instead of the associated tracks for high p_{T} objects. Both versions of the $p_{\mathrm{T}}^{\mathrm{miss}}$ show good performance for a number of Run 1 event topologies, with Monte Carlo simulations that describe the data well, excellent resolution stability with pileup, and a good measurement of the scale of the event.

References

[1] Aad G *et al.* (ATLAS Collaboration) 2008 *JINST* **3** S08003
[2] Aad G *et al.* (ATLAS Collaboration) 2013 *ATLAS-CONF-2013-082*
[3] Cacciari M, Salam G P and Soyez G 2008 *JHEP* **0804** 063 (*Preprint* 0802.1189)
[4] Aad G *et al.* (ATLAS Collaboration) 2013 *Eur.Phys.J.* **C73** 2304 (*Preprint* 1112.6426)
[5] Aad G *et al.* (ATLAS Collaboration) 2014 (*Preprint* 1406.0076)
[6] Aad G *et al.* (ATLAS Collaboration) 2009 (*Preprint* 0901.0512)
[7] Cacciari M and Salam G P 2008 *Phys.Lett.* **B659** 119–126 (*Preprint* 0707.1378)
[8] Mangano M L, Moretti M, Piccinini F, Pittau R and Polosa A D 2003 *JHEP* **0307** 001 (*Preprint* hep-ph/0206293)
[9] Butterworth J, Forshaw J R and Seymour M 1996 *Z.Phys.* **C72** 637–646 (*Preprint* hep-ph/9601371)
[10] Frixione S and Webber B R 2002 *JHEP* **0206** 029 (*Preprint* hep-ph/0204244)
[11] Corcella G, Knowles I, Marchesini G, Moretti S, Odagiri K *et al.* 2001 *JHEP* **0101** 010 (*Preprint* hep-ph/0011363)

Journal of Physics: Conference Series **645** (2015) 012013 doi:10.1088/1742-6596/645/1/012013

Using a Classical Gluon Cascade to study the Equilibration of a Gluon-Plasma

Lucas McConnell
Department of Physics, University of Cape Town, South Africa
E-mail: lmhmcconnell@gmail.com

Abstract. Using a classical gluon cascade, we study the thermalisation of a gluon-plasma in a homogeneous box by considering the time evolution of the entropy, and in particular how the thermalisation time depends on the strong coupling α_s. We then partition the volume into cells with a linearly increasing temperature gradient in one direction, and homogeneous/isotropic in the the other two directions. We allow the gluons to stream in one direction in order to study how they then evolve spatially. We examine cases with and without collisions. We study the entropy as well as the flow-velocity in the z-direction and find that the system initially has a flow which dissipates over time as the gluons become distributed homogeneously throughout the box.

1. Introduction

Normal hadronic matter is composed of quarks in either three-quark or quark-antiquark bound states which are held together by the strong force. These quarks are said to be confined, as their strong attraction prevents them from being observed outside of these bound hadron states. However, under extremely high densities, it is possible for the quarks to become free of this strong attraction and become de-confined. When this happens the resultant state of matter is referred to as a Quark-Gluon Plasma (QGP). A QGP is a phase of Quantum Chromodynamics (QCD) that consists of free quarks and gluons in a state of matter that resembles (but is distinct from) a gas-like plasma. It is believed that up to a few milliseconds after the Big Bang that the universe was in a QGP state and thus its study can provide interesting insights into our earliest universe. In contrast to a gas-like plasma, the QGP behaves like a near-ideal Fermi liquid. The creation of the QGP is the main subject of the heavy ion collisions at the Relativistic Heavy Ion Collider (RHIC) and at the Large Hadron Collider (LHC) at CERN. To recreate the conditions of the primordial universe, these powerful accelerators collide head-on heavy ions such as gold or lead. In these heavy ion collisions, hundreds of protons and neutrons that make up two such nuclei smash into each other at energies of upwards of a few trillion electronvolts each. This forms a miniscule fireball in which everything “melts” into a QGP.

The QGP can be described by the Boltzmann equation which gives a microscopic picture of particle systems. Thus in order to describe the QGP we solve the relatavistic Boltzmann equation

$$\left(\frac{\partial}{\partial t} + \frac{\mathbf{p}}{E}\nabla\right) f(\mathbf{x}, \mathbf{p}) = C_{22}, \tag{1}$$

where $f(\mathbf{x}, \mathbf{p})$ is the particle density in phase space, and C_{22} is the collision term. Unfortunately,

Journal of Physics: Conference Series **645** (2015) 012013 doi:10.1088/1742-6596/645/1/012013

the Boltzmann equation has no analytic solution and must be solved using numerical methods. One such method is to solve it via Monte Carlo cascade simulations. We implement here such a simulation of a simplified picture of the QGP, and use it to study the thermalisation of a gluon only plasma in a volume. We will consider only elastic $gg \to gg$ collisions and will treat the gluons as classical Boltzmann particles. The cascade itself involves two distinct stochastic processes: to determine whether or not two gluons actually collide and to determine exactly how they collide.

2. Background

2.1. Scattering in the CoM Frame

It is often easier to work out the specifics of $2 \to 2$ scattering in the centre of mass (CoM) frame, which is defined as the frame wherein we have $\mathbf{p}_1 + \mathbf{p}_2 = 0$. We use a Lorentz boost to transform from the Lab frame to the CoM frame, work out the details of the scattering, before performing the inverse Lorentz boost back to the Lab frame. In the CoM Frame (quantities in the CoM Frame are primed), our pair of colliding massless gluons will have equal energies $E_1' = E_2' = \sqrt{\mathbf{p}_1' \cdot \mathbf{p}_1'}$ and thus equal three-momenta but in opposite directions $\mathbf{p}_1' = -\mathbf{p}_2'$. Mandelstam s is then

$$\begin{aligned} s &= (P_1' + P_2')^2 \\ &= (|\mathbf{p}_1'| + |\mathbf{p}_2'|)^2 \\ &= (2|\mathbf{p}_1'|)^2 = 4E_1'^2 = 4E_2'^2 \end{aligned} \tag{2}$$

Similarly Mandelstam t is

$$\begin{aligned} t &= (P_1' - P_3')^2 = 0 - (\mathbf{p}_1' - \mathbf{p}_3')^2 \\ &= -2\mathbf{p}_1'^2 + 2\mathbf{p}_1'\mathbf{p}_3' \cos\theta' \\ &= -2E_1'^2(1 - \cos\theta') = -\frac{s}{2}(1 - \cos\theta') \end{aligned} \tag{3}$$

Where the zero in the first line comes from conservation of four-momentum and θ describes to the scattering angle in the CoM frame. Note from Eq. (3) that since $\cos\theta' \in [-1, 1]$, that $t \in [-s, 0]$.

The three-momentum transfer is given by $\mathbf{q}^2 = (\mathbf{p}_1' - \mathbf{p}_3')^2$. Thus we have that $t = -\mathbf{q}^2$. The three-momenta in the CoM frame for a gluon i involved in the scattering is given by the Lorentz boost [3],

$$\mathbf{p}_i' = \mathbf{p}_i + \frac{\gamma - 1}{\mathbf{v}_{CoM}^2}(\mathbf{v}_{CoM} \cdot \mathbf{p}_i)\mathbf{v}_{CoM} - \gamma \mathbf{v}_{CoM} E_i \tag{4}$$

$$, \tag{5}$$

where the boost velocity and γ factor of the CoM are

$$\mathbf{v}_{CoM} = \frac{\mathbf{p}_1 + \mathbf{p}_2}{|\mathbf{p}_1| + |\mathbf{p}_2|} \tag{6}$$

$$\gamma = \sqrt{\frac{1}{1 - \mathbf{v}^2}} \tag{7}$$

Our axes are defined by the units vectors

$$\hat{\mathbf{x}}' = \frac{\mathbf{p}_i'}{|\mathbf{p}_i'|}, \tag{8}$$

$$\hat{\mathbf{y}}' = N^{-1}[\mathbf{v}_{com} - (\mathbf{v}_{com} \cdot \hat{\mathbf{x}}')\hat{\mathbf{x}}']. \tag{9}$$

$$\hat{\mathbf{z}}' = \hat{\mathbf{x}}' \times \hat{\mathbf{y}}'. \tag{10}$$

Journal of Physics: Conference Series **645** (2015) 012013 doi:10.1088/1742-6596/645/1/012013

Where N is a normalisation factor. Then the scattered momenta in terms of the scattering angles in the CoM frame are thus given by [2]

$$\mathbf{p}'_{i_{scattered}} = p'_i \left(\cos\theta'\hat{\mathbf{x}}' + \sin\theta'\sin\phi'\hat{\mathbf{y}}' + \sin\theta'\cos\phi'\hat{\mathbf{z}}'\right). \tag{11}$$

The scattering angles are determined via inverse transform sampling before performing the inverse Lorentz boost to get back to the Lab frame.

2.2. Stochatically determining if the Gluons Scatter

We derive the collision probability directly from the collision term of the Boltzmann equation (Eq. 1). The particle density in phase space is

$$f(x,p) = \frac{\Delta N}{\frac{1}{(2\pi)^3}\Delta^3 x \Delta^3 p_i}, \tag{12}$$

while (neglecting quantum effects) the collision term C_{22} is given by [1]

$$\begin{aligned} C_{22} = \frac{1}{2E_1}\int \frac{d^3p_2}{(2\pi)^3 2E_2}\frac{1}{\nu}\int \frac{d^3p'_1}{(2\pi)^3 2E'_1}\frac{d^3p'_2}{(2\pi)^3 2E'_2} f'_1 f'_2 \times |M_{1'2'\to 12}|^2 (2\pi)^4 \delta^4(p'_1 + p'_2 - p_1 - p_2) \\ -\frac{1}{2E_1}\times\int \frac{d^3p_2}{(2\pi)^3 2E_2}\frac{1}{\nu}\int \frac{d^3p'_1}{(2\pi)^3 E'_1}\frac{d^3p'_2}{(2\pi)^3 2E'_2} f_1 f_2 \times |M_{12\to 1'2'}|^2 (2\pi)^4 \delta^{(4)}(p_1 + p_2 - p'_1 - p'_2), \end{aligned} \tag{13}$$

where ν is a factor related to double counting and $|M_{1'2'\to 12}|^2$ is the leading order pQCD interaction matrix element squared.

If we assume the two particles occupy a spatial volume $\Delta^3\mathbf{x}$ and have momenta in the range $(\mathbf{p}_1, \mathbf{p}_1 + \Delta^3 p_1)$ and $(\mathbf{p}_2, \mathbf{p}_2 + \Delta^3 p_2)$, then the collision rate per unit phase space for incoming particles p_1 and p_2 with $\Delta^3 p_1$ and $\Delta^3 p_2$ will be given by

$$\frac{\Delta N_{coll}}{\Delta t \frac{1}{(2\pi)^3}\Delta^3 x \Delta^3 p_1} = \frac{1}{2E_1}\frac{\Delta^3 p_2}{(2\pi)^3 2E_2} f_1 f_2 \times \frac{1}{2}\int \frac{d^3p'_1}{(2\pi)^3 2E'_1}\frac{d^3p'_2}{(2\pi)^3 2E'_2}|M_{12\to 1'2'}|^2 (2\pi)^4 \delta^{(4)}(p_1 + p_2 - p'_1 - p'_2). \tag{14}$$

We can re-write Eq. (14) as [1]

$$\frac{\Delta N^{22}_{coll}}{\Delta t \frac{1}{(2\pi)^3}\Delta^3 x \Delta^3 p_1} = \frac{1}{2E_1}\frac{\Delta^3 p_2}{(2\pi)^3 2E_2} f_1 f_2 \times 2s \times \sigma_{22}. \tag{15}$$

Then using Eq. (12), we obtain an expression for the absolute collision probability in a unit box $\Delta^3 x$ and unit time Δt,

$$P_{22} = \frac{\Delta N^{22}_{coll}}{\Delta N_1 \Delta N_2} = v_{rel}\sigma_{22}\frac{\Delta t}{\Delta^3 x}, \tag{16}$$

where $v_{rel} = \frac{s}{E_1 E_2}$, and s is the invariant mass of the particle pair.

We will need the scattering cross-section in order to obtain the collision probabilities. For elastic gluon scattering, the differential cross-section is [1]

$$\frac{d\sigma^{gg\to gg}}{dt} = \frac{9\pi\alpha_s}{2s^2}\left(3 - \frac{tu}{s^2} - \frac{su}{t^2} - \frac{st}{u^2}\right), \tag{17}$$

where s, t, and u are the Mandelstam invariants.

Journal of Physics: Conference Series **645** (2015) 012013 doi:10.1088/1742-6596/645/1/012013

The collision probability requires a cross-section. The cross-section is large when t is small, corresponding to a small scattering angle which are overwhelmingly probable. The differential cross-section is then approximated by [1]

$$\frac{d\sigma}{dt} \approx \frac{9\pi\alpha_s}{t^2}. \tag{18}$$

To avoid infrared divergence, a Debye mass term is included to account for screening.

$$\frac{d\sigma}{dt} \approx \frac{9\pi\alpha_s^2}{(t - m_D^2)^2} \tag{19}$$

Integrating this over all t (i.e. from $-s$ to 0) then gives a regularised finite total cross-section

$$\sigma(s) = 9\pi\alpha_s^2 \frac{s}{m_D^2(s + m_D^2)}. \tag{20}$$

3. Simulation

With the background set out in the previous section, we are now equipped to simulate a gluon-plasma in a volume. Two separate configurations were examined. The first focuses on gluons colliding within a homogeneous box where the equilibrium temperature, number of gluons, strong coupling constant as well as the actual time period for which the gluon-plasma evolves are all variable parameters. Figure 1 shows an example of studying the thermalisation of the homogeneous box via the time evolution of its entropy. Through algebraic manipulation, we are able to fix the equilibrium temperature of the system through our choice of initial parameters. The system was simulated for a variety of initial phase space distributions including a delta function. After a characteristic time, called the thermalisation time, the system was observed to reach a state of dynamical equilibrium. We extracted the thermalisation time by modelling the difference of the entropy at each time step from its analytical equilibrium value as an exponential decay.

In the second configuration we partition the volume into cells and allow the gluons to stream in one direction. We consider two cases: one in which the gluons free-stream and the other where they are allowed to collide but only with other gluons in their cell. Figures 2 and 3 shows example plots of the flow-velocity in the z-direction (extracted using the energy-momentum tensor) as a function of time for free-streaming and including collisions respectively.

4. Conclusion

We found that the homogenous box system thermalises faster with increasing equilibrium temperature and particle number. We were then able to simulate the system for a variety of values for α_s and thus study how the thermalisation time is proportional to the variation of the QCD coupling constant. The thermalisation time was qualitatively shown to be inversely proportional to α_s.

We then added a spatial aspect to the model by dividing the volume into cells and allowing the gluons to stream in one direction. We examined a case where the gluons did not collide amongst themselves (free-streaming) and a case where they did. We implemented a linearly increasing temperature gradient for the cells and sampled the momenta of the gluons from a Boltzmann distribution. The entropy for both cases was seen to increase, with the free-streaming case gaining entropy from mixing, while the case which included collisions gained entropy from both mixing and thermalisation. We also studied how the gluons spatially distribute themselves throughout the volume. In both cases the gluon-plasma eventually reached a point where the gluons were distributed homogeneously throughout the volume. Though in the latter case, the

Journal of Physics: Conference Series **645** (2015) 012013 doi:10.1088/1742-6596/645/1/012013

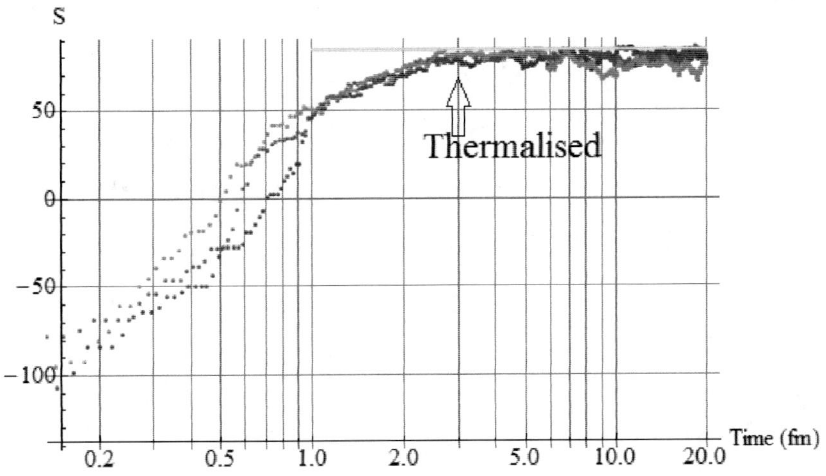

Figure 1: Time evolution of the entropy (initial phase space distribution is a delta function) for a plasma of 128 gluons with $T_{eq} = 300$ MeV, $\alpha_s = 0.4$,and $dp = 0.2$. The solid line plotted alongside is the analytic value for the entropy at equilibrium.

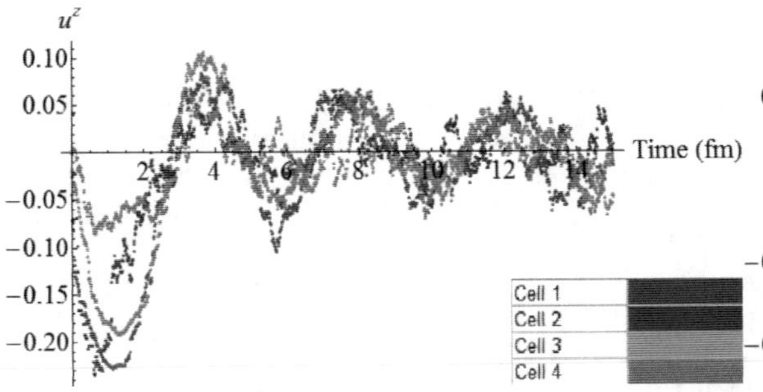

Figure 2: z-direction flow-velocity (free-streaming)

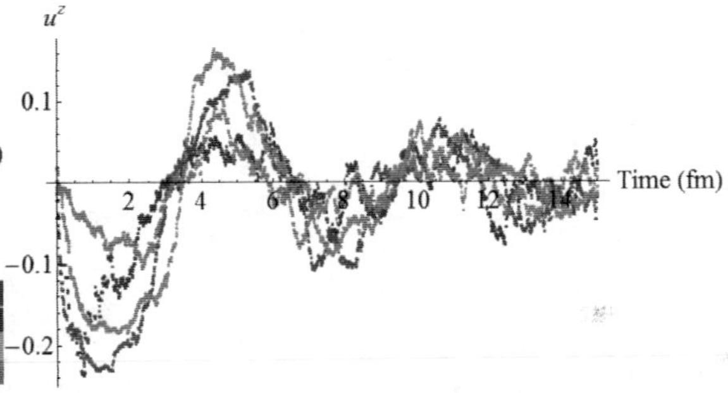

Figure 3: z-direction flow-velocity (collisions incl.)

end situation was noted to be a dynamic equilibrium which constantly fluctuated about its equilibrium values. We also derived from the energy-momentum tensor, the flow-velocity in the z-direction. We found that in both cases, the system initially had a flow-velocity in the negative z-direction which, after a characteristic time, reversed direction but also decreased in magnitude. This behaviour repeated until the flow-velocity went to zero (apart from fluctuations).

5. References

[1] Z. Xu and C. Greiner, Phys. Rev. C **71** (2005) 064901 .
[2] B. Svetitsky, Phys. Rev. D **37** (1988) 2484.
[3] J.D. Jackson. *Classical Electrodynamics 3rd Edition.* John Wiley and Sons, Inc, 1998.
[4] L.D. Landau and E.M. Lifshitz. *Statistical Physics, Third Edition, Part 1: Volume 5 (Course of Theoretical Physics, Volume 5).* Butterworth-Heinemann, 1980
[5] J. Kapusta and C. Gale. *Finite-temperature Field Theory*, Cambridge University Press, 2006
[6] H.C. Ohanian and R. Ruffini. *Gravitation and Spacetime*, WW Norton and Company, 1994

Performance of missing transverse momentum reconstruction in ATLAS studied in proton-proton collisions in 2012 at 8 TeV

Luis March

School of Physics, University of the Witwatersrand, Johannesburg 2050, South Africa

E-mail: lmarch@cern.ch

Abstract. The missing transverse energy plays a really important role in reconstructing events produced at hadron colliders. Undetectable particles, such as neutrinos, pass through the matter with a negligible probability of interaction. Hence, no direct evidence of them can be measured in a general purpose detector, as ATLAS. However, the total momenta in the transverse plane to the beam axis has to be conserved and computed. In particular, it is used in searches for the Standard Model Higgs boson channels, such as: $H \to WW$, $H \to ZZ$ and $H \to \tau\tau$. The benefit of using this conservation law is that an energy imbalance may signal the presence of such undetectable particles. Therefore, it becomes also a powerful tool for new physics searches at the Large Hadron Collider, such as Supersymmetry and Extra Dimensions. The performance of the missing transverse momentum reconstruction in the ATLAS detector is evaluated using data collected in 2012 in proton-proton collisions at a centre-of-mass energy of 8 TeV. An optimised reconstruction of missing transverse momentum is used and the effects arising from additional proton-proton interactions superimposed on the hard physics process are suppressed with various methods. Results are shown for a data sample corresponding to an integrated luminosity of about 20 fb^{-1} and for events with different topologies with or without a genuine missing transverse momentum due to undetected particles.

1. Introduction

In a hadron collider event the missing transverse momentum is defined as the momentum imbalance in the plane transverse to the beam axis, where momentum conservation is relevant. Such an imbalance may signal the presence of undetectable particles, such as neutrinos or new weakly-interacting particles. The vector momentum imbalance in the transverse plane is obtained from the negative vector sum of the momenta of all particles detected in a proton-proton (pp) collision and is denoted as missing transverse momentum, E_T^{miss}.

An optimised reconstruction and calibration of E_T^{miss} [1] was developed by the ATLAS Collaboration. The E_T^{miss} measurement is significantly affected by the contributions of additional pp collisions superimposed on the hard physics process, referred to as pile-up in the following, so methods were developed to suppress such contributions [2]. This article describes the performance, in terms of resolution, response and tails, of the reconstructed E_T^{miss} after pile-up suppression.

The event samples used to assess the quality of the E_T^{miss} reconstruction are events with leptonically decaying W and Z bosons. These test the detector capability in the reconstruction

Journal of Physics: Conference Series **645** (2015) 012014 doi:10.1088/1742-6596/645/1/012014

of different physics objects and the methods of pile-up suppression. The E_T^{miss} performance is studied in both data and Monte Carlo simulation and comparisons are made. In simulated events, the E_T^{miss} is calculated from all non-interacting particles in the event, including neutrinos from heavy flavour decay, and is referred to as true E_T^{miss} ($E_T^{miss,True}$) in the following.

This article is organised as follows. Section 2 shows the data samples and event selection used for this performance study, followed by Section 3, where the Monte Carlo simulation samples are also shown. In Section 4, the E_T^{miss} reconstruction is described and, Section 5 lists the different methods for pile-up suppression. Section 6 shows the results for the E_T^{miss} performance studies and finally, in Section 7, the conclusions are summarized.

2. Data samples and event selection

During 2012, proton-proton (pp) collisions at a centre-of-mass energy of 8 TeV were recorded with stable proton beams and nominal ATLAS magnetic field condidtions. Only data with a fully functioning calorimeter, inner detector and muon spectrometer are analysed.

The data sets used correspond to a total integrated luminosity of approximately 20 fb^{-1}, collected in 2012 with a bunch crossing interval (bunch spacing) of 50 ns. The mean number of interactions per bunch crossing was about 20.7, reaching values up to 35 at the beginning of a fill during the 2012 LHC running period.

Selection criteria applied for the study of events with leptonically decaying W and Z bosons are described in these references [1, 2, 3].

3. Monte Carlo simulation samples

Monte Carlo (MC) samples of $Z \to ll$ and $W \to l\nu$ production are generated with the next-to-leading (NLO) order POWHEG [4] model, with the final state partons showered by the PYTHIA8 [5, 6] program and the ATLAS AU2 tune [7].

Additional inelastic pp interactions, known as pile-up interactions, are generated using the PYTHIA8 program with the ATLAS MC12 A2M tune [7] and the MSTW08 leading order PDF [8].

The GEANT4 software toolkit [9] within the ATLAS simulation framework simulates the propagation of the generated particles through the ATLAS detector and their interactions with the detector material.

The same trigger and event selection criteria used for $Z \to ll$ and $W \to l\nu$ data are also applied to the simulated events.

4. E_T^{miss} reconstruction

The E_T^{miss} reconstruction [1] uses energy deposits in the calorimeters and muons reconstructed in the muon spectrometer. Tracks are added to recover the contribution from low-p_T particles which are missed in the calorimeters.

The E_T^{miss} calculation uses reconstructed and calibrated physics objects. Calorimeter energy deposits are associated with a reconstructed and identified high-p_T parent object in a specific order: electrons, photons, hadronically decaying τ-leptons, jets and finally muons. The E_T^{miss} is calculated as follows:

$$E_{x(y)}^{miss} = E_{x(y)}^{miss,e} + E_{x(y)}^{miss,\gamma} + E_{x(y)}^{miss,\tau} + E_{x(y)}^{miss,jets} + E_{x(y)}^{miss,SoftTerm} + E_{x(y)}^{miss,\mu} \quad (1)$$

where each term is calculated as the negative sum of the calibrated reconstructed objects, projected onto the x and y directions. Only jets with calibrated p_T greater than 20 GeV are used to calculate the jet term in Equation 1. The soft term is calculated from topoclusters and tracks not associated to high-p_T objects (i.e. from unassociated topoclusters and tracks).

Journal of Physics: Conference Series **645** (2015) 012014 doi:10.1088/1742-6596/645/1/012014

The total transverse energy in the calorimeters, $\sum E_T$, which includes also the unassociated low-p_T tracks used in the soft term, is an important quantity to parameterise and understand the E_T^{miss} performance. It is defined as the scalar sum:

$$\sum E_T = \sum E_T^e + \sum E_T^\gamma + \sum E_T^\tau + \sum E_T^{jets} + \sum E_T^{SoftTerm}$$

which is the scalar sum of the transverse energy of reconstructed and calibrated objects and of the soft term according to the scheme described above for E_T^{miss}. The total transverse energy in the event is obtained by summing the p_T of muons and the $\sum E_T$ in the calorimeters:

$$\sum E_T(event) = \sum E_T + \sum p_T^\mu \qquad (2)$$

5. Methods for pile-up suppression in E_T^{miss}

In Reference [2], it was shown that a clear deterioration of the performance is observed when the average number of pile-up interactions per event increases. Methods to suppress pile-up are therefore needed, which can restore the E_T^{miss} resolution to values more similar to the ones observed in the absence of pile-up.

All E_T^{miss} terms in Equations 1 and 2 are affected by pile-up, but the terms which are most affected are the jets and soft term, because the pile-up largely produces hadronic energy and they are reconstructed from larger regions in the calorimeters. Methods for the suppression of pile-up in these terms are summarized in this section.

5.1. Pile-up suppression in the E_T^{miss} jet term based on tracks

Pile-up not only distorts the energy reconstructed in jets but can also create additional jets. To further suppress the jets originating from pile-up, a cut is applied based on the Jet Vertex Fraction (JVF) [10], i.e. the fraction of momenta of tracks matched to the jet which are associated with the hard scattering vertex. JVF is defined as:

$$JVF = \sum_{tracks_{jets},PV} p_T / \sum_{tracks_{jets}} p_T$$

where the sums are taken over the tracks matched to the jet and PV denotes the tracks associated to the Primary Vertex (PV). Only tracks with $p_T > 400$ MeV and passing further quality criteria relating to impact parameters and number of hits in different tracking sub-detectors are used to make primary vertices.

5.2. Pile-up suppression in the E_T^{miss} soft term based on tracks

The pile-up largely affects the soft term. Since the $E_T^{miss,SoftTerm}$ can have an important contribution to the momentum balance in the event, completely neglecting its contribution in the E_T^{miss} reconstruction gives a poorer performance [2]. Two different methods for suppressing the pile-up in the soft term are described in the following, one based on the use of tracks and the other one based on the jet area method.

Tracks provide an excellent method for pile-up suppression, since they can be associated with the primary vertex from the hard scattering collision. It is calculated, in a similar way as JVF, as:

$$STVF = \sum_{tracks_{SoftTerm},PV} p_T / \sum_{tracks_{SoftTerm}} p_T$$

where the sums are taken over the tracks unmatched to physics objects and PV denotes the tracks associated to the primary vertex. The $E_T^{miss,SoftTerm}$ is multiplied by the STVF factor and the E_T^{miss} calculated, with this corrected soft term, is named STVF.

5.3. Pile-up suppression in the E_T^{miss} soft term using the jet area method

The event transverse momentum density (ρ) is used to determine the contribution due to pile-up in the jet area, which is subtracted from each jet: $p_T^{jet,corr} = p_T^{jet} - \rho A^{jet}$. Two different $E_T^{miss,SoftTerm}$ calculations are considered here and the E_T^{miss} is then recalculated using each of them. The two methods differ only in their calculation of the ρ. One is named Extrapolated Jet Area Method and the other named Jet Area Filtered.

6. Study of E_T^{miss} performance in different event topologies

6.1. Characterization of samples used for the study of E_T^{miss} performance

The E_T^{miss} performance depends on the event topology: presence of leptons, jet activity, etc. The values of quantities relevant for E_T^{miss} studies in the MC samples used in this note are shown in Figure 1. In Figure 1 the reconstructed $\sum E_T$ before and after pile-up suppression is

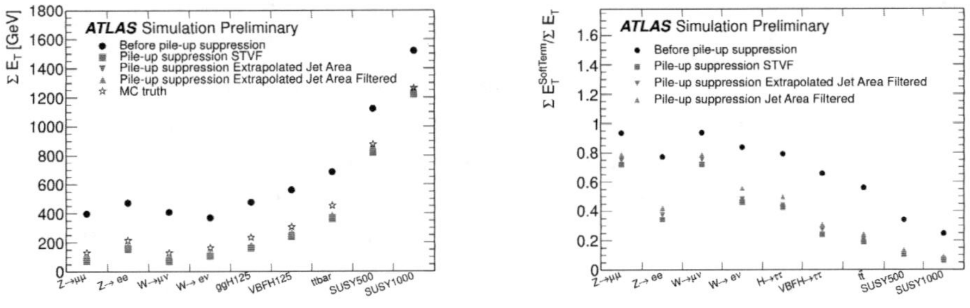

Figure 1. Average value of $\sum E_T$ (left) and ratio $\sum E_T^{miss,SoftTerm} / \sum E_T$ (right) in the MC samples used. The values before and after pile-up suppression with different methods are shown. Plots obtained from this reference [3].

compared with the true values. As can be seen, the $\sum E_T$ is strongly suppressed by all the pile-up suppression methods and it is closer to the true value. Figure 1 (right) shows the importance of the soft term, in the different samples. The average value of $E_T^{miss,SoftTerm}$ is not much different in the various samples but its contribution is dominant in Z samples and important in W events, while it becomes less important in events with higher jet multiplicity, where the $E_T^{miss,jets}$ is dominant, as can be seen from the distribution of the ratio of $\sum E_T^{SoftTerm} / \sum E_T$.

6.2. Comparison of E_T^{miss} distributions in $Z \rightarrow \mu\mu$: Data and MC simulation

The distributions of E_T^{miss} for data are shown in Figure 2 for $Z \rightarrow \mu\mu$ events. The MC simulation expectations, from $Z \rightarrow \mu\mu$ events and from the dominant backgrounds, are superimposed. Each MC sample is weighted with its corresponding cross-section and then the total MC expectation is normalized to the number of events in data. A good agreement between data and MC simulation is observed in the E_T^{miss} distribution, both before and after pile-up suppression.

6.3. Study of E_T^{miss} resolution

A first study of the E_T^{miss} resolution is performed using the ratio:

$$R = RMS(E_T^{miss}/E_T^{miss,True}) / < E_T^{miss}/E_T^{miss,True} >$$

Figure 3 shows the ratio R for two samples. Before pile-up suppression, R decreases with increasing $E_T^{miss,True}$ and is of the order of 0.1 for larger $E_T^{miss,True}$ values in both samples. After

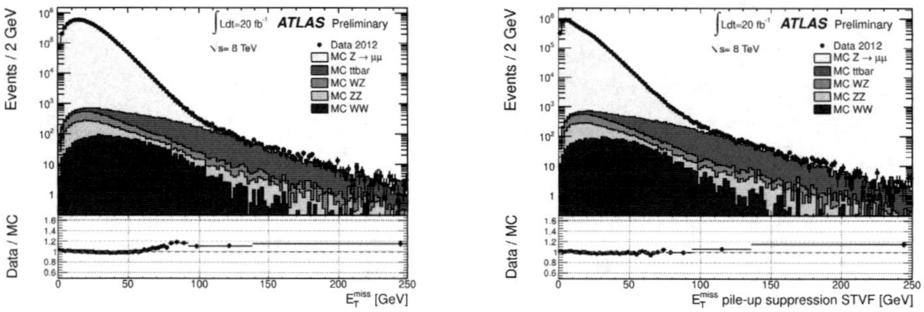

Figure 2. Distribution of E_T^{miss} measured in data samples $Z \to \mu\mu$ events before (left) and after pile-up suppression with the STVF method (right). The lower part shows the ratio data/MC. Plots obtained from this reference [3].

pile-up suppression, in $W \to l\nu$ events, there is a reduction of R in the region of $E_T^{miss,True} < 40$ GeV, while the improvement of R is smaller for larger $E_T^{miss,True}$ values. In the low $E_T^{miss,True}$ region there are some concurrent effects: the region is mostly populated by events without jets, so the reduction of R indicates that the pile-up suppression methods in the soft term improve the $E_T^{miss,True}$ resolution.

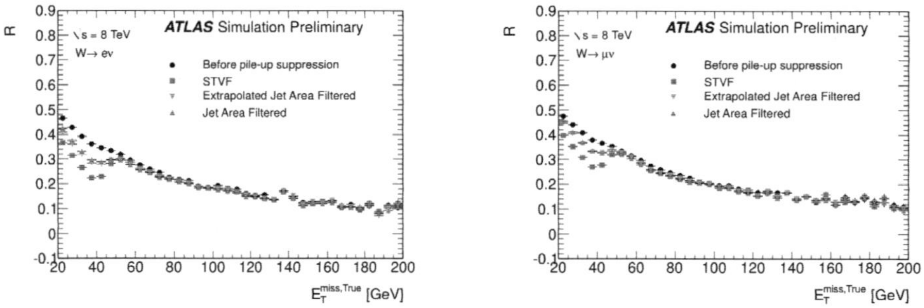

Figure 3. Resolution (R) as function of $E_T^{miss,True}$ in MC $W \to e\nu$ (left) and $W \to \mu\nu$ (right). Plots obtained from this reference [3].

6.4. Study of E_T^{miss} linearity

The E_T^{miss} linearity is defined as the mean value of the ratio: $(E_T^{miss} - E_T^{miss,True})/E_T^{miss,True}$. The mean value of this ratio is expected to be zero if E_T^{miss} is reconstructed at the correct scale. The linearity for the different MC samples considered is shown in Figure 4.

For both $W \to l\nu$ samples a positive bias is observed for low $E_T^{miss,True}$ values which is due to the finite resolution of the E_T^{miss} measurement. For larger $E_T^{miss,True}$ values, the bias is within 5%. Differences in the linearity before and after pile-up suppression are visible. In particular, a larger non-linearity is observed for STVF, because of the strong pile-up suppression mainly in events without jets.

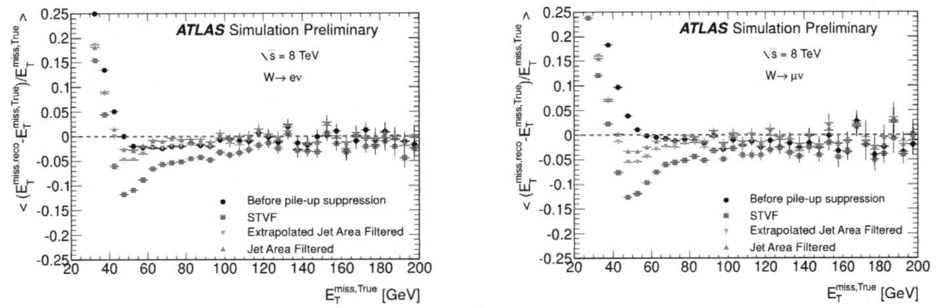

Figure 4. E_T^{miss} linearity in $W \rightarrow e\nu$ (left) and $W \rightarrow \mu\nu$ (right) MC events vs $E_T^{miss,True}$. Plots obtained from this reference [3].

7. Conclusions

The missing transverse momentum E_T^{miss} performance has been studied in events with different topologies in proton-proton collisions at a centre-of-mass energy of 8 TeV recorded with the ATLAS detector in 2012.

The value of E_T^{miss} is calculated from calibrated reconstructed objects and from the unmatched topological clusters and tracks ($E_T^{miss,SoftTerm}$). Several methods for pile-up suppression in the soft term are described, based on the use of tracks (STVF method) or on the jet area method.

The Monte Carlo simulation describes the data in general rather well. Some discrepancy in data-MC comparison is observed after pile-up suppression in the $E_T^{miss,SoftTerm}$ and in the contribution from jets, due to corrections applied for pile-up suppression.

The E_T^{miss} resolution improves after pile-up suppression in events where the contribution of the soft term is important and it becomes closer to that observed in the absence of pile-up, mainly the STVF.

The linearity of the E_T^{miss} measurement is studied in MC simulation as afuntion of the true E_T^{miss}. Except for the bias observed at small true E_T^{miss} values (visible up to 40 GeV), due to the finite E_T^{miss} resolution, the linearity is better than 5%.

References

[1] ATLAS Collaboration, 2012 Performance of missing transverse momentum reconstruction in proton-proton collisions at 7 TeV in ATLAS. *Eur.Phys.J.* C **72**

[2] ATLAS Collaboration, 2012 Performance of Missing Transverse Momentum Reconstruction in ATLAS with 2011 Proton-Proton collisions at $\sqrt{s}$ =7 TeV. *ATLAS-CONF-2012-101*

[3] ATLAS Collaboration, 2013 Performance of Missing Transverse Momentum Reconstruction in ATLAS studied in Proton-Proton Collisions recorded in 2012 at $\sqrt{s}$ =8 TeV. *ATLAS-CONF-2013-082*

[4] Nason P, 2004 A new method for combining NLO QCD with shower Monte Carlo algorithms. *JHEP* **0411** 040

[5] Sjostrand T, Mrenna S and Skands PZ, 2008 A Brief Introduction to PYTHIA 8.1. *Comput.Phys.Commun.* **178** 852-867

[6] Corke R and Sjostrand T, 2011 Interleaved Parton Showers and Tuning Prospects. *JHEP* **032** 1103

[7] ATLAS Collaboration, 2012 Summary of ATLAS Pythia 8 tunes. *ATLAS-PHYS-PUB-2012-003*

[8] Martin A et al, 2009 Parton distributions for the LHC. *Eur.Phys.J.* **C63**

[9] GEANT4 Collaboration, 2003 GEANT4: A simulation toolkit. *Nucl.Instrum.Meth.* **A506**

[10] ATLAS Collaboration, 2009 Expected performance of the ATLAS experiment - detector, trigger and physics (Jet and E_T^{miss} chapter). *tech.rep.*

Measurement of the Higgs Boson Transverse Momentum in the Di-photon Channel with the ATLAS detector in Run 1

Robert Graham Reed

School of Physics, University of the Witwatersrand , Johannesburg 2050, South Africa

E-mail: robert.reed@cern.ch

Abstract. The Standard Model (SM) of particle physics, with the discovery of the Higgs boson, is a model of the known fundamental particles and their interactions. The data taken in the 2012 run was then compared to the Monte Carlo and an excess has been found in the Higgs transverse momentum in the di-photon and ZZ decay channels. A possible explanation is a beyond the SM scalar boson is being produced which would then decay into a dark matter particle and a Higgs boson that looks like the current SM. This dark matter particle would provide the Higgs with excess momentum which may account for the discrepancy observed. A first attempt at modelling the production of the heavier than the SM Higgs (or scalar boson) showed that as the centre of mass energies increase the production cross-section of the scalar boson increased faster than the SM Higgs boson. This indicates that if the hypothesis is true then we should expect greater Higgs boson productions during the 2015 run at higher centre of mass energies. A better understanding of the observed excess is needed before any further conclusions can be made.

1. Introduction

On July 4th 2012 both the ATLAS and CMS Collaborations announced the discovery of a Higgs-like boson with a mass of 125 GeV [2,3]. Over a year later in subsequent papers [4, 5] this boson was observed to be the Standard Model (SM) Higgs Boson with a significance greater than 7σ as seen in Fig. 1. The Higgs boson is predicted by the Standard Model as a consequence to spontaneous symmetry breaking which gives mass to elementary particles. The discovery has been a highlight for the LHC. The new particle is a tantalizing new tool for beyond the standard model searches.

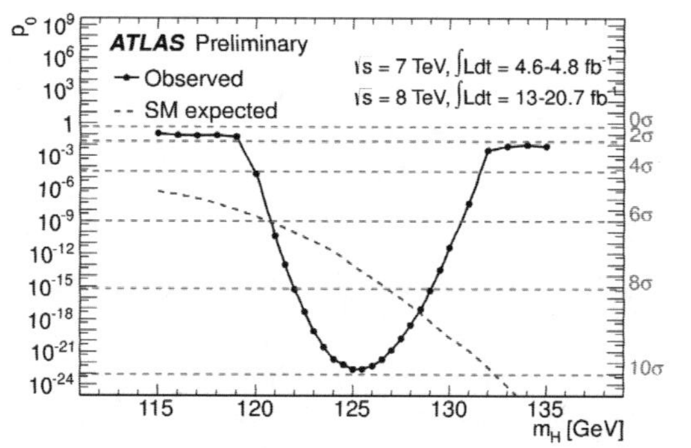

Figure 1: Combined search showing the local probability [1].

Journal of Physics: Conference Series **645** (2015) 012015 doi:10.1088/1742-6596/645/1/012015

2. The ATLAS Detector

The ATLAS detector is one of two main general purpose detector at the LHC. The ATLAS detector is different in many respects to the CMS detector most noticeable is the air-core superconducting magnets that produce the toroidal field for the muon spectrometer. The detector is cylindrical in shape and has a forward-backward symmetry. The detector consists of concentric sub detectors. Starting from the inside we have the inner tracker and silicon microstrip detector located inside the transition radiation tracker which provide tracking in the pdeudorapidity range $|\eta| < 2.5$, electromagnetic calorimeter and lead liquid-argon sampling device is divided into one barrel ($|\eta| < 1.475$)) and two end cap regions ($1.375 < |\eta| < 3.2$), the hadron calorimeter is also divided into a barrel ($|\eta| < 0.8$) and extended barrel ($0.8 < |\eta| < 1.7$) both of which consist of steel and plastic scintillators. Each sub detector is designed to identify different particles, their energies and trajectories. After combining all sub detector readings and apply statistical techniques to extract the signals one can piece together a snippet of what occurred during a collision event.

3. Event Selections

An additional challenge is faced once the raw data has been reduced to "interesting" events. How we identify objects is a complicated process that involves many different techniques and statistical tools. In this proceedings I will not discuss in detail the object identification process but rather the event selection for the Higgs to di-photons channel only. The details of the object identification for the Higgs to di-photon channel can be found in Ref. [6].

To distinguish between electrons and photons the reconstruction algorithm looks for possible matches between energy clusters and tracks reconstructed in the inner detector and extrapolated to the calorimeter. Well-reconstructed tracks matched to clusters are classified as electron candidates while clusters without matching tracks are classified as unconverted photon candidates. These photon candidates are then calibrated using statistical tools that have been optimised for this task. When selecting photons its required to have two photons in the difucial region defined by $|\eta| < 2.37$ and must exclude the gap region defined by $1.37 < |\eta| < 1.56$. The two photons with the highest energy are then considered. The ratio of the transverse energy (E_T) to the invariant mass of the two photons (m_{yy}) for the leading and subleading photons must be greater than 0.35 and 0.25 respectively.

The measurement of the correct production vertex is important to accurately measure the invariant mass of the diphoton system. The vertex is selected from the reconstructed collision vertices using a neural-network algorithm. The efficiency of finding the correct vertex in simulated Higgs to diphotons is approximately 83 % which improves as the transverse momentum increases.

4. Higgs Kinematic Results

The higgs boson transverse momentum is studied as it is linked to the kinematics of the Higgs boson production. In this section I will show results of the Higgs boson transverse momentum in the diphoton channel for both ATLAS and CMS as well as the four lepton channel from ATLAS. Figure 2 shows the results of the Higgs boson transverse momentum in the diphoton decay channel at the ATLAS detector. The log scale on the y-axis must be noted. In this plot we have greater resolution in bin number due to the larger signal yield compared to the four lepton channel. We see a similar structure as before with low p_T being over shot by the MC, the region $20 < p_T < 100GeV$ is undershot by the MC and high p_T is agreeable with predictions. Due to the higher signal yield the excess is more prevalent in the Higgs to di-photons channel. Figure 3a) shows the transverse momentum of the Higgs boson in a decay to four leptons at the ATLAS detector.

The first thing we notice is the large error bars on the data. Assuming these error bars reduce with more data we can see a structure in the p_T that is not predicted by the Monte Carlo (MC). The data is overshot for low p_T, undershot for range of approximately $20 < p_T < 60GeV$ and matching fine for high p_T. Figure 3b) is the same result but from the CMS detector. Similar results are seen with an excess in the same region as before. Each plot shows an excess in the mid p_T range. How significant is this excess is the next natural question to ask. As a preliminary run a Kolmogorov-Smirnov test is performed on each individual plot and an overall KS test statistic is calculated.

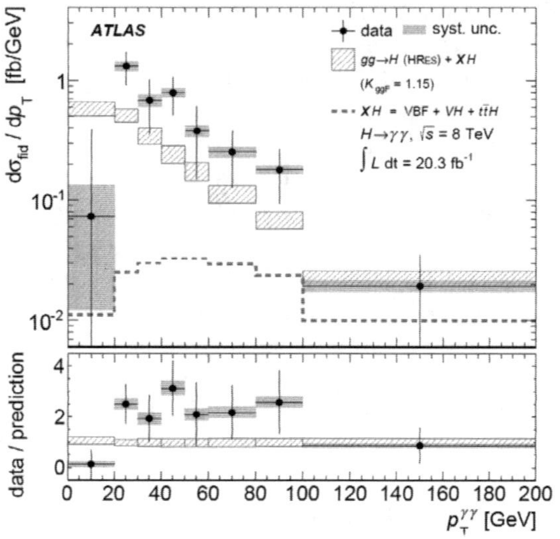

Figure 2: Transverse momentum of the Higgs boson in the diphoton channel.

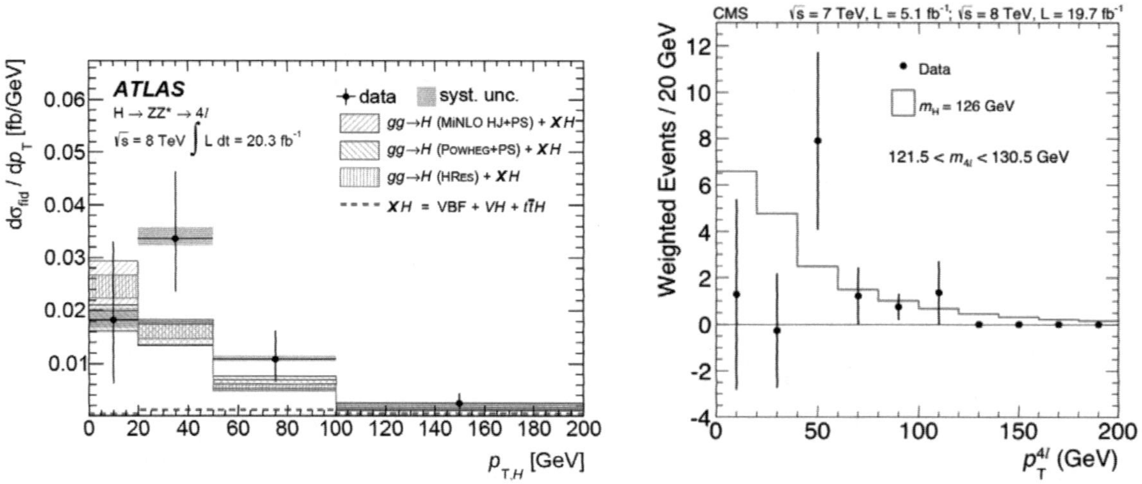

(a) (b)

Figure 3: Transverse momentum of the Higgs in the four lepton channel. a) ATLAS and b) CMS.

4.1. Kolmogorov Test

The two sample Kolmogorov-Smirnov test is used to test how likely two samples are produced from the same set. It is based on the maximum difference between an empirical and a hypothetical cumulative distribution. The test statistic is given by Eq. 1.

$$D_{n,n'} = sup_x \left| F_{1,n}(x) - F_{2,n'}(x) \right| \tag{1}$$

Table 1: Kolmogorov-Smirnov Test Results

Channel	**KS Value**
$H \to 4l$ ATLAS	0.635814
$H \to 4l$ CMS	0.301351
$H \to \gamma\gamma$ ATLAS	0.029377
TOTAL	$\mathbf{0.00563 \approx 2.8\sigma}$

where, $D_{n,n'}$ is the Kolmogorov-Smirnov test statistic KS, sup_x is the supremum or maximum of the set, $F_{1,n}(x)$ *and* $F_{2,n'}(x)$ are the empirical distribution functions. The distribution functions are calculated from the total number of events after taking into account the luminosity and differential cross sections for each graph independently. Table 1 shows the results obtained for each plot shown in Fig. 3 and Fig. 2. The most noticeable is the diphoton case which has a small test statistic. Taking the product of all three cases gives the overall test statistic which is equivalent to approximately 2.8σ.

5. Discussion and Conclusions

The results of the Kolmogorov-Smirnov test indicate a significant excess from two different decay channels which is seen in ATLAS and CMS. An excess of 3σ may or may not stay after the next data taking session in mid 2015. It is possible that this excess is merely a statistical fluctuation but a more exciting alternative is a beyond the SM explanation. The extra p_T could be explain by the production of an extra scalar boson that decays into a Higgs boson and dark matter. The Higgs boson recoils off the dark matter which can produce an excess in p_T that we pbserve.. Once the detector is online and new measurements are studied will we be able to better answer this question. The fact that there is currently an excess, even if it may be statistical fluctuations, warrants an investigation.

6. Acknowledgements

The University of the Witwatersrand has facilitated in the efforts by purchasing the required hardware to allow hands on development for this work and for other big science applications. The National Research Foundation (NRF) for the bursary they provided. The SA-CERN program provided financial assistance for research visits to CERN. I would also like to thank the School of Physics, the Faculty of Science and the Research Office at the University of the Witwatersrand.

References

[1] The ATLAS Collaboration 2013 *48th Rencontres de Moriond on QCD and High Energy Interactions* 25 URL `https://cds.cern.ch/record/1528170`
[2] The ATLAS Collaboration 2012 *Physics Letters B* **716** 1–29 URL `http://www.sciencedirect.com/science/article/pii/S037026931200857X`
[3] The CMS Collaboration 2012 *Physics Letters, Section B: Nuclear, Elementary Particle and High-Energy Physics* **716** 30–61
[4] The CMS Collaboration 2014 *The European Physical Journal C* **74** 3076 URL `http://link.springer.com/10.1140/epjc/s10052-014-3076-z`
[5] The ATLAS Collaboration 2012 **111803** 1–19 URL `http://dx.doi.org/10.1103/PhysRevLett.108.111803`
[6] The ATLAS Collaboration 2014 **112015** 1–44 URL `http://inspirehep.net/record/1310835`

Measurement of Higgs boson properties using the diphoton decay with the ATLAS detector

X. Ruan

School of Physics, University of the Witwatersrand, Johannesburg 2050, South Africa

E-mail: `xifeng.ruan@cern.ch`

Abstract. The Standard Model Higgs boson like particle was observed in July 2012 by both the ATLAS and CMS collaboration at the LHC. The measurements of the Higgs boson properties are performed in Higgs boson decaying into two photons channel after the discovery. Higgs boson production by the gluon-gluon fusion, the vector boson fusion and in association with a W or Z boson or the top-quark pair are measured in this final state. The multivariate analysis method is applied to extract the Higgs boson in the vector boson fusion enriched category to enhance the signal significance. The couplings of each production modes are measured using $20.3\,\mathrm{fb}^{-1}$ 2012 data taken at center-of-mass energy $\sqrt{s}=8\,\mathrm{TeV}$ and $4.5\,\mathrm{fb}^{-1}$ 2011 data taken at $\sqrt{s}=7\,\mathrm{TeV}$ in ATLAS. No significant deviation from the Standard Model prediction is found. The fiducial and differential cross sections of Higgs boson are measured using 2012 data in this final state as well. The distribution of several kinematic variables of the two photons and jets are studied. The results are compared with several theoretical predictions.

Keywords: ATLAS; Higgs boson; Diphoton; Multivariate analysis; Differential; Fiducial

1. Introduction

The Standard Model (SM) Higgs-boson-like particle was observed in 2012 by ATLAS and CMS [1] [2]. After the discovery, the measurements of Higgs properties became important. The Higgs boson coupling strengths in gluon-gluon fusion production, vector boson fusion production (VBF), vector boson associated production (VH) and top production (ttH) were measured. The Higgs boson mass was measured as well. Based on the measured Higgs boson event yields in each bin of the kinematic distribution, the differential cross sections and fiducial cross sections were also measured.

The ATLAS detector [3] is one of the general purpose detectors at the LHC. Its four sub-detectors: the inner detector, the electromagnetic calorimeter, the hadronic calorimeter and the muon spectrometer provide the precise track and energy measurements. Electron, muon, photon and jets are reconstructed in the detector. The measurements were performed by combining the $20.3\,\mathrm{fb}^{-1}$ 8 TeV data taken in 2012 and $4.5\,\mathrm{fb}^{-1}$ 7 TeV data taken from 2011. The differential and fiducial cross section measurements were only using 2012 8 TeV data. The photon energy was re-calibrated in 2013 including implementing the multivariate analysis (MVA) method to improve the identification performance. In the coupling measurement, the multivariate analysis method was also used to improve the sensitivity. The unfolding technique was implemented to extract the Higgs boson kinematic distribution from data to the particle level. The results were compared with several Monte Carlo (MC) prediction using higher order QCD calculation.

Journal of Physics: Conference Series **645** (2015) 012016 doi:10.1088/1742-6596/645/1/012016

2. The Event Selection

In the Higgs boson decaying into the diphoton channel, events are required to have at least two photons. The events must pass the trigger criteria that the transverse energy (E_T) should be larger than 35 GeV and 25 GeV for the leading (highest E_T) and subleading photons, respectively. The photons are restricted to be within the fiducial calorimeter region of $|\eta| < 2.37$ [1] and to exclude the transition region between the barrel and the end-cap calorimeters, $1.37 < |\eta| < 1.56$. The relative p_T: $p_T/m_{\gamma\gamma}$ of the leading and subleading photon should be larger than 0.35 GeV and 0.25 GeV, respectively, in which the $m_{\gamma\gamma}$ is the invariant mass of the two photons. The photons are well identified and isolated. The jets are defined to have $E_T > 25\,\text{GeV}$ (30 GeV) for $|\eta| < 2.4$ ($2.4 < |\eta| < 4.5$).

For the coupling measurement, all of the events which have two photons are categorised into several regions dedicated for different production. It includes the ttH, VH, VBF enriched category. The remaining events are categorised into the bins defined by different conversion configuration or the calorimeter regions. The events in the VBF region are decided using MVA method. The boosted decision tree (BDT) [4] is used. Six kinematic variables are chosen as the inputs to discriminate the signal and background. Each event is given a BDT score and categorised into VBF tight, VBF loose and the other categories, as shown in the left of Fig. 1. The six variables are not correlated to the diphoton invariant mass. They are chosen as below:

- p_{Tt}: Diphoton p_T projected perpendicular to the diphoton thrust axis.
- $\Delta\eta_{jj}$: Pseudo-rapidity separation between the leading two jets. One of the biggest signature of the VBF production mode. The VBF events tend to have two forward jets so that they have larger jet η separation.
- $\eta^\star$: $\eta_{\gamma\gamma} - \langle\eta_{jj}\rangle$.
- $\Delta R^{\gamma j}_{\text{min}}$: Minimum ΔR between either the leading/subleading jet and the leading/subleading photon, in which $\Delta\text{R} = \sqrt{\Delta\phi^2 + \Delta\eta^2}$.
- m_{jj}: invariant mass of the leading and subleading jets.
- $\Delta\Phi_{\gamma\gamma jj}$: Azimuthal angle between the diphoton and the two jets.

The training has used POWHEG [5] MC at next-to-next-to-leading order (NNLO) QCD and next-to-leading order (NLO) electro-weak (EW) corrections. The diphoton background sample for the training is simulated by Sherpa [6].

3. The Coupling and Mass Results

In each category, a signal plus background fit is performed on the $m_{\gamma\gamma}$ shape simultaneously. The couplings are parameterised using the signal strength: μ_{ggF}, μ_{VBF}, μ_{VH} and $\mu_{t\bar{t}H}$, defined as the ratio of the measure event yields over the predicted event yields. The results are shown in the right of Fig. 1 [7].

$$\begin{aligned}\mu_{ggF} &= 1.32 \pm 0.38\\ \mu_{VBF} &= 0.8 \pm 0.7\\ \mu_{WH} &= 1.0 \pm 1.6\\ \mu_{ZH} &= 0.1^{+3.71}_{-0.1}\\ \mu_{t\bar{t}H} &= 1.6^{+2.7}_{-1.8}\end{aligned}$$

The mass measurement is performed with the different categorisation. The diphoton events are divided into 10 categories labeled by low or high p_{Tt}, central, forward or transition region,

[1] η is the pseudo-rapidity, defined as $\eta = -\ln[\tan(\theta/2)]$, θ is the angle between the particle momentum and the beam axis.

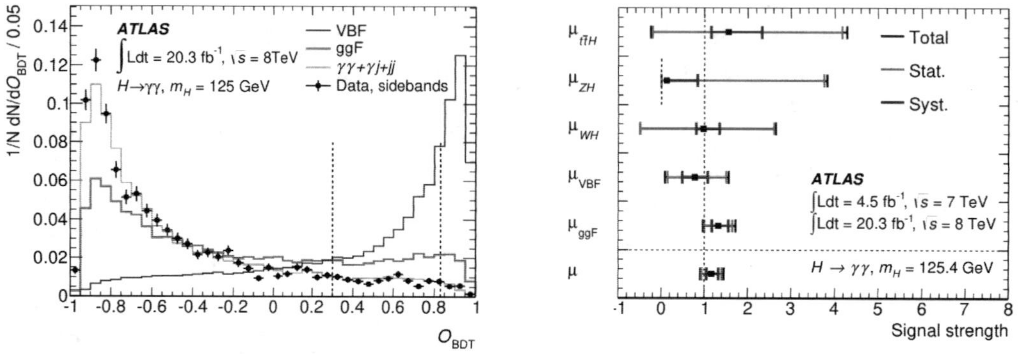

Figure 1. Left: the BDT output of the signal, background and data. The vertical line shows the boundary of categorisations. Right: the measured signal strength in the different Higgs boson production modes [7].

converted or non-converted photons. The unbinned maximum likelihood fit is performed simultaneously in the categories and the 7 TeV and 8 TeV data. The largest systematic uncertainty is the photon energy scale uncertainty. By combining the results from the Higgs boson decaying into four lepton channel, the result is obtained [8]:

$$m_H = 125.36 \pm 0.37(stat) \pm 0.18(syst)\,\text{GeV}$$

4. The Differential and Fiducial Cross Sections

The Higgs boson event yields are measured in each bin of the differential distribution by fitting the invariant mass spectrum. The distributions include the jet activity, Higgs boson kinematics, VBF sensitive variables and spin-CP variable. The variables are listed as below:

- Higgs boson kinematics: $p^T_{\gamma\gamma}$, $|y_{\gamma\gamma}|$
- Jet activity: Jet multiplicity N_{jets}, $p_T^{j_1}$, $|y_{j_1}|$, $p_T^{j_2}$, the scalar sum of jet transverse momenta, H_T
- Spin-CP sensitive variables: $\Delta\phi_{jj}$, $|cos\theta^{\star}|$
- VBF-sensitive variables: $|\Delta\phi_{\gamma\gamma,jj}|$, $|\Delta y_{jj}|$

The signal yields are corrected for the effects of detector inefficiency and resolution. The results are obtained by using the bin-by-bin unfolding correction factors derived from MC samples. The results are compared with several theoretical predictions describing the Higgs boson+jets activities. The HRes 2.2 [9] calculation is used to provide the inclusive kinematics of the diphoton system via gluon fusion. HRes is accurate to NNLO+NNLL in QCD. The BLPTW (soft-collinear effective theory, NNLO+NNLL 0-jet + NLO+NLL 1-jet cross sections), JetVHeto and MINLO [10] provide predictions for events associated with jets. No significant mismodelling is observed according to the MC prediction. However a small deviation in Higgs p_T distribution is observed, as shown in the left of Fig. 2. The 2015 new data will help to exclude or verify this phenomenon. The jet multiplicity distribution is shown in the right of Fig. 2.

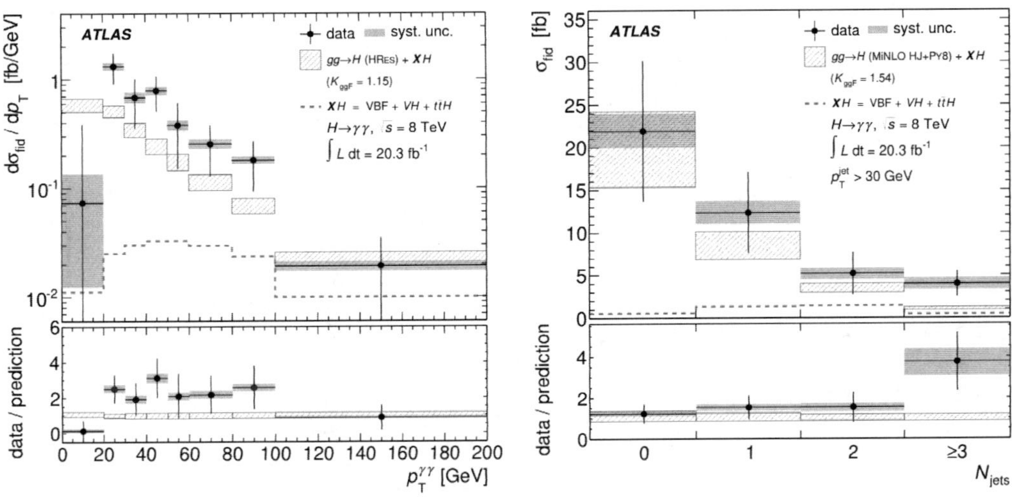

Figure 2. Left: the differential cross section of p_T^H measured in two photon final state. Right: the jet multiplicity distribution. The jet p_T threshold is 30 GeV [11].

The fiducial cross section is also measured in the fiducial range $p_{T\gamma,leading(subleading)}/m_{\gamma\gamma} >$ 0.35(0.25) and 105 GeV$< m_{\gamma\gamma} <$ 160 GeV:

$$43.2 \pm 9.4(stat)^{+3.2}_{-2.9}(syst) \pm 1.2(lumi) \text{ fb}$$

The cross sections are measured in several fiducial regions as well. They are measured in the inclusive jet bins and the VBF enhanced region. The limits on the cross sections at the 95% confidence level are obtained in the diphoton production with leptons region and the high E_T^{miss} region. The results are compared with several theoretical predictions, as shown in Fig. 3.

5. Conclusions

The Higgs boson mass and coupling were measured after the recalibration of photon energy in ATLAS in 2014. The differential and fiducial cross section were measured as well. No significant deviation from the SM prediction was observed except a small distortion on the Higgs boson p_T distribution. The 2015 Run II data taking is important to verify the measured result, to measure the Higgs boson p_T distribution and to discover the new physics with higher centre mass energy. The multivariate analysis were performed to improve the diphoton analysis. This method will also benefit the 2015 analysis.

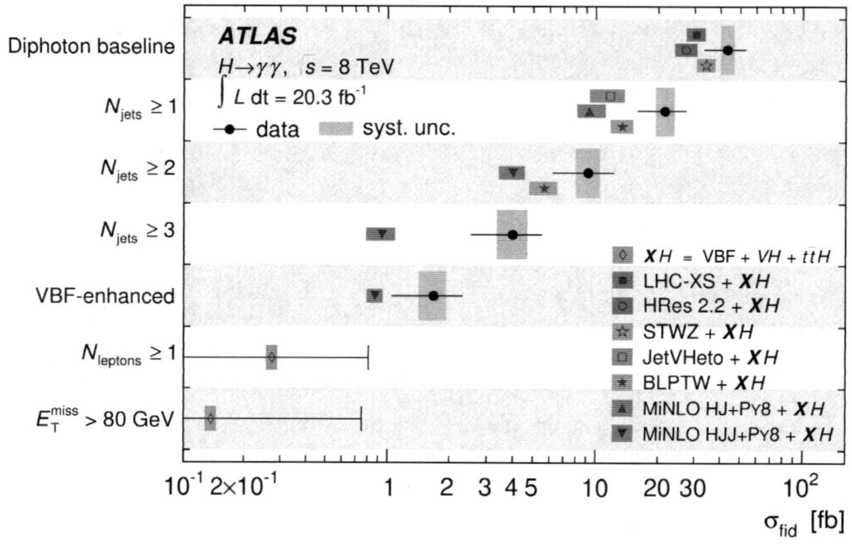

Figure 3. The measured cross sections and cross-section limits for $pp \to H$ in the seven fiducial regions. The intervals on the vertical axis each represent one of these fiducial regions. The data are shown as filled (black) circles. The error bar on each measured cross section represents the total uncertainty in the measurement, with the systematic uncertainty shown as dark grey rectangles. The error bar on each cross-section limit is shown at the 95% confidence level. The width of each theoretical prediction represents the total uncertainty in that prediction. All regions include the SM prediction arising from VBF, VH and ttH, which are collectively labelled as XH [11].

References

[1] ATLAS Collaboration, Phys.Lett. B716 (2012) 1, arXiv:1207.7214[hep-ex]
[2] CMS Collaboration, Phys.Lett. B716 (2012) 30, arXiv:1207.7235 [hep-ex]
[3] ATLAS Collaboration, The ATLAS experiment at the CERN large hadron collider, J.Instrum., 3 (2008)
[4] A. Hoecker, P. Speckmayer, J. Stelzer, J. Therhaag, E. von Toerne, and H. Voss, TMVA 4 (Toolkit for multivariate data analysis with ROOT) Users Guide. 2009
[5] S. Alioli, P. Nason, C. Oleari, and E. Re, JHEP 0904 (2009) 002, arXiv:0812.0578 [hep-ph], P. Nason and C. Oleari, JHEP 1002 (2010) 037, arXiv:0911.5299 [hep-ph]
[6] T. Gleisberg, S. Hoeche, F. Krauss, M. Schonherr, S. Schumann, et al., JHEP 0902 (2009) 007, arXiv:0811.4622 [hep-ph]
[7] ATLAS Collaboration, Phys. Rev. D 90, 112015.
[8] ATLAS Collaboration, Phys. Rev. D 90, 052004.
[9] D. de Florian, G. Ferrera, M. Grazzini, and D. Tommasini, JHEP 06 (2012) 132, [arXiv:1203.6321], M. Grazzini and H. Sargsyan, JHEP 09 (2013) 129, [arXiv:1306.4581].
[10] R. Boughezal, X. Liu, F. Petriello, F. J. Tackmann, and J. R. Walsh, Phys. Rev. D89 (2014) 074044, [arXiv:1312.4535], A. Banfi, P. F. Monni, G. P. Salam, and G. Zanderighi, Phys. Rev. Lett. 109 (2012) 202001,[arXiv:1206.4998], K. Hamilton, P. Nason, and G. Zanderighi, JHEP 10 (2012) 155, [arXiv:1206.3572].
[11] ATLAS Collaboration, JHEP 1409 (2014) 112, arXiv:1407.4222 [hep-ex]

Journal of Physics: Conference Series **645** (2015) 012017 doi:10.1088/1742-6596/645/1/012017

Hypothesising Dark Matter Models in pp Collisions

Stefan von Buddenbrock

School of Physics, University of the Witwatersrand, Wits 2050, South Africa

E-mail: stef.von.b@cern.ch

Abstract. Recent analysis of Large Hadron Collider (LHC) Run 1 data has shown an apparent mismodelling is Higgs p_T from ATLAS $\gamma\gamma$ and 4ℓ decay channels, as well as the CMS 4ℓ decay channel.This has been referred to as the p_T crisis, and it is postulated that it can be explained by the Higgs being produced in combination with a dark matter particle. A minimal Z' is modelled as a dark matter mediator and shown to have cross sections too low to satisfy current experimental constraints. The viability of a two-Higgs-doublet model providing a form of dark matter is then argued to be a viable solution to an excess of double Higgs production events from Run 1 LHC data, and an experimental analysis done by ATLAS is given as motivation. I conclude by noting that these analyses will be enhanced by data from Run 2 of the LHC.

1. Apparent Mismodelling in the Standard Model

It goes without saying that the discovery of the Standard Model (SM) Higgs boson in 2012 [1, 2] has sparked the interest of the global particle physics community. With a large amount of confidence that the discovered particle is indeed the Higgs boson predicted by the SM, physicists are now asking questions about whether the Higgs boson carries with it some signature of physics beyond the Standard Model (BSM).

Experimental observations are, at the moment, the only resource by which we can infer the existence of BSM physics. By measuring the properties of Higgs production and comparing the data to what we would expect from the SM, we can look at instances of what we call *mismodelling* – that is, the SM may not reproduce what we observe in experiment. It has become apparent that the ATLAS and CMS pp collision results from Run 1 of the LHC are indeed exhibiting a sort of mismodelling.

In particular, I will consider the so-called p_T crisis. Recent analysis of LHC Run 1 data has shown that there is a mismodelling of Higgs transverse momentum (p_T) spectra. Experimentally obtained p_T spectra indicate that the data is assuming a different structure from what is expected in the SM. This can be seen from the ATLAS $h \to \gamma\gamma$ decay results [3] (shown in figure 1), the ATLAS $h \to ZZ \to 4\ell$ decay results [4], and the CMS $h \to ZZ \to 4\ell$ decay results [5]. My primary interest is the apparent excess of p_T in these spectra. An excess of p_T might imply that the Higgs boson produced in the LHC is recoiling off of a particle which, along with its decay products, does not interact with the detector at all. I will refer to this type of particle as a *dark matter* particle. The purpose of this study is to test the viability of dark matter production at the LHC being responsible for the p_T crisis.

The existence of dark matter was first postulated by astrophysical measurements. Nowadays, we regard dark matter as making up about 24% of the universe's energy content. As a particle physicist, however, one can think of dark matter merely as a massive particle which does not

Journal of Physics: Conference Series **645** (2015) 012017 doi:10.1088/1742-6596/645/1/012017

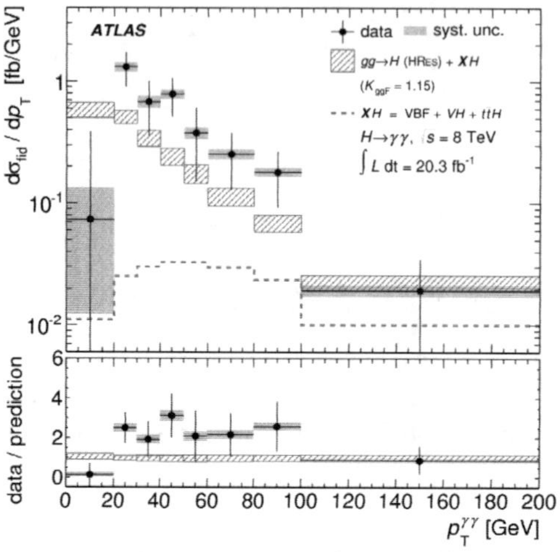

Figure 1. The p_T spectra arising from performing analysis on the ATLAS Run 1 diphoton decay channel data and extracting fiducial cross sections of the processes [3].

interact electromagnetically. Therefore, any model which may be proposed for the existence of dark matter is currently believed to satisfy a set of two experimental constraints. The first set of constraints is one which we can attribute to its particle nature, and these constraints are set by the fundamental principles of quantum field theory as well as particle detector data. The second set of constraints comes from astrophysical observations. For instance, a dark matter candidate (or candidates) must be able to explain the dark matter density in the universe [6]

$$\Omega_{\text{DM},0}h^2 = \frac{\rho h^2}{\rho_0} = 0.1196 \pm 0.0031,$$

where h here is the scale factor of the universe and $\Omega_{\text{DM},0}$ is the energy density of dark matter in the universe. Thus, the study of dark matter is inherently interesting and difficult due to the fact that it links particle physics with cosmology.

2. Z′ Dark Matter

By Occam's Razor, we note that it is often wisest to consider the simplest hypothesis when attempting to explain phenomena. In the case of dark matter, the simplest model is one which extends the gauge symmetry of the SM by including an extra $U(1)$ symmetry. An extra $U(1)$ predicts the addition of an extra particle, which is usually denoted by Z'. The model which I have considered is the *minimal* Z' model, which has a corresponding conserved *baryon minus lepton* charge, denoted by $B - L$ for short [7].

In pp collisions we would expect the leading order of Z' production to come from tree level reactions such the diagram shown in figure 2. This is where the virtue of the Higgs boson can be seen in detecting dark matter. Since the Z' is considered an *invisible* particle, and will not interact electromagnetically, the best chance of detecting such a particle will come by virtue of its mass. Since the Z' is considered to be a massive particle, it will couple to the Higgs, and we can therefore use the Higgs boson as a *portal* in studying possible dark matter candidates.

In order to survey this possibility, a `MadGraph` [8] model was generated from a `FeynRules` package implemented by Lorenzo Basso [7]. This computational model was created using the minimal Z' physical model described above. The model allows for the variation of the Z' mass as well as its associated gauge couplings. In this work I determined the effect of varying the mass of the Z', with the couplings at their default values, $g_1' = 0.2$ and $\tilde{g} = -0.1$.

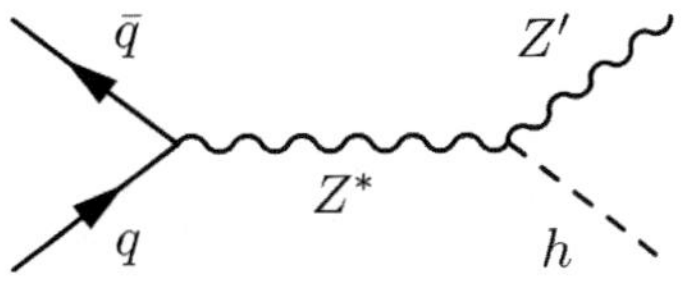

Figure 2. Leading order Z' production from pp collisions.

The process which was modelled has the form

$$q\bar{q} \to Z \to Z'H,$$

the Feynman diagram of which is given in figure 2. For all simulations we used a center of mass energy of $\sqrt{s} = 14\text{TeV}$. By varying the mass of the Z' between values of 0.05TeV and 1.5TeV I was able to plot the cross section as a function of $M_{Z'}$, which is shown in figure 3.

We note from figure 3 that there is a deducible constraint on the mass of the Z', should we seek to detect the particle. The production cross section is only substantial at masses smaller than $\sim$ 100GeV. This implies that if we are to detect a Z' there is very little chance that we could detect a Z' with a mass in the TeV scale. This puts the idea of Z' dark matter in a rather unconvincing light, since it can be argued that dark matter mediated by a Z' could only be plausible if the Z' mass is in the TeV scale [9].

3. Two-Higgs-Doublet Models

The SM of particle physics is built by incorporating and studying electroweak symmetry breaking (EWSB) – that is, the spontaneous breaking of the $SU(2)_L \times U(1)_Y$ gauge symmetry. After EWSB, we are left with four massive bosonic states: the Z boson, the W^+ and W^- bosons, and the Higgs boson h. This is because we impose the existence of a complex scalar doublet (which has four degrees of freedom), the Higgs doublet ϕ. The SM Higgs potential has the form

$$V_H(\phi) = \mu^2\phi^\dagger\phi + \lambda\left(\phi^\dagger\phi\right)^2.$$

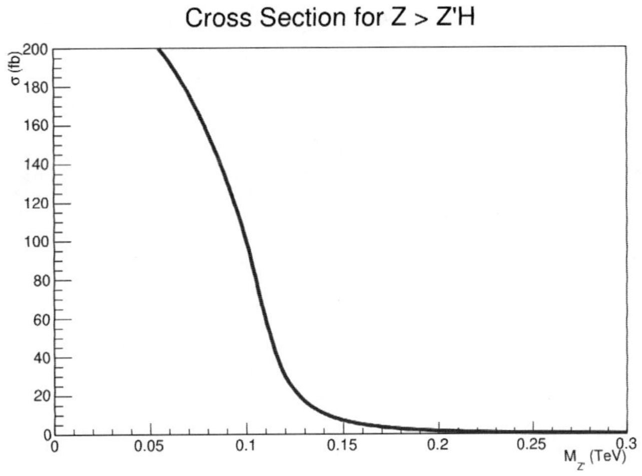

Figure 3. The cross section for tree level Z' production reactions in association with a Higgs boson, as shown in figure 2. The mass of the Z' was varied, while keeping its associated couplings constant.

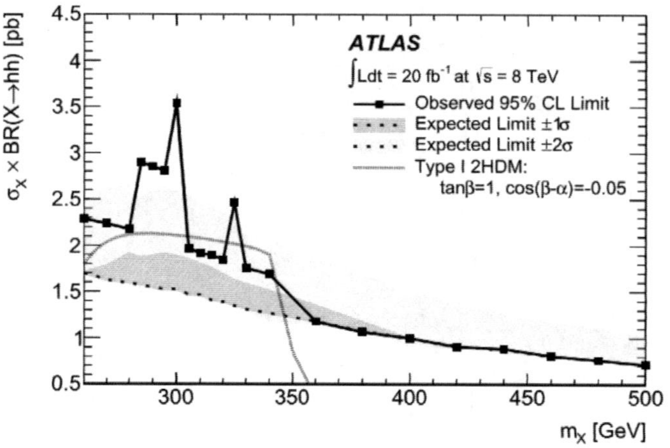

Figure 4. A 95% CL upper limit on the cross section times branching ratio of a resonance X decaying to a Higgs pair as a function of the resonance mass [11].

The experimental discovery of the Higgs boson is well explained by this structure.

However, what has been observed experimentally could be explained equally as well by imposing two Higgs doublets in the construction. If we call these doublets ϕ_1 and ϕ_2, then the general renormalisable Higgs potential can be written down as [10]

$$V_H(\phi_1,\phi_2) = m_{11}^2|\phi_1|^2 + m_{22}^2|\phi_2|^2 - m_{12}^2\phi_1^\dagger\phi_2 - \left(m_{12}^2\right)^*\phi_2^\dagger\phi_1 + \frac{1}{2}\lambda_1|\phi_1|^4 + \frac{1}{2}\lambda_2|\phi_2|^4$$
$$+ \lambda_3|\phi_1|^2|\phi_2|^2 + \lambda_4\left(\phi_1^\dagger\phi_2\right)\left(\phi_2^\dagger\phi_1\right) + \frac{1}{2}\lambda_5\left(\phi_1^\dagger\phi_2\right)^2 + \frac{1}{2}\lambda_5^*\left(\phi_2^\dagger\phi_1\right)^2.$$

Any model which incorporates two Higgs doublets in this way is called a Two-Higgs-Doublet Model (2HDM). The spontaneous symmetry breaking of a 2HDM requires us to introduce a different vacuum expectation value (VEV) for the second Higgs doublet, v_2, in addition to the normal VEV we are familiar with, v_1. The physical VEV is then given by

$$v^2 = {v_1}^2 + {v_2}^2,$$

and the two VEVs are related by an angle β such that

$$\tan\beta = \frac{v_2}{v_1}.$$

The physical implication of introducing a second Higgs doublet is that four new massive physical states are introduced. These are a heavy CP even Higgs boson H (which is heavier than the SM Higgs boson by convention), a CP odd Higgs boson A, and two charged Higgs boson H^+ and H^-. The two neutral CP even Higgs bosons are related by a mixing angle α. One might treat these new states (or at least their potential decay products) as dark matter particles.

While I have done no work on modelling a 2HDM state as a dark matter particle, it should be noted that the existence of these states is currently being taken very seriously in certain ATLAS analyses. One such analysis is, which I will consider, that of an excess of double Higgs production from Run 1 LHC data. Using the SM, one shouldn't expect to find a significant contribution resulting from pp collisions with a final state of two Higgs bosons. However, Run 1 LHC data seems to indicate an excess of double Higgs production events. One explanation for this is that a heavy resonance X is decaying into a Higgs pair.

Figure 4 shows the 95% CL limit while modelling a resonance decay with the variation of its mass. This analysis was conducted by studying the $\gamma\gamma b\bar{b}$ decay channel of the Higgs pair [11]. It seems evident that Run 1 data seems to favour a resonance mass of about 300GeV. We could model this resonance as being the heavy Higgs state predicted by a 2HDM, since a heavy Higgs could decay into two SM Higgs bosons. Should one choose to use a 2HDM to study the process, one would expect contributions from the diagrams shown in figure 5.

Although it seems that this excess is due to a resonance decaying, it should be noted that there is a possibility of it being due to a non-resonance effect as well. There is also, of course, the possibility that the excess is merely a statistical fluctuation, but only more data will shed light on this possibility.

4. Concluding Remarks

The p_T crisis resulting from Run 1 LHC data presents us with the possibility to explore BSM physics. The fact that an excess in Higgs p_T is seen alludes to the idea that the Higgs boson is being produced in combination with a massive and invisible particle which may be thought of as a dark matter particle. While the possibility of a Z' mediating dark matter seems to be far from feasible with current constraints, there is a recognisable possibility of finding a state predicted by a 2HDM, although more data is needed to start work on classifying whether or not it could count as a discovery. The p_T crisis itself could merely be a statistical fluctuation. As such, I conclude by saying that many of the results and ideas presented are in need of clarification which only more experimental data could provide, and these analyses will benefit from the data to be accrued by Run 2 of the LHC.

Acknowledgements

I acknowledge Prof. Bruce Mellado and Prof. Alan Cornell for their supervision of this work. This work is supported by the National Research Foundation of South Africa, as well as the Claude Leon Foundation. I would also like to acknowledge Dr. Mukesh Kumar, Dr. Yann Mambrini and Dr. Jahred Adelman for their helpful input and discussions.

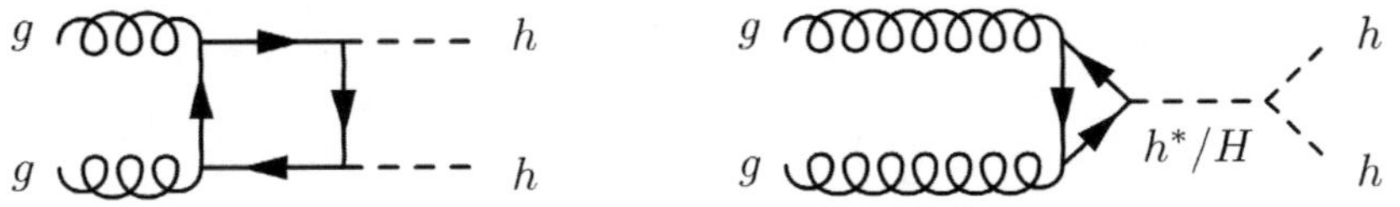

Figure 5. The leading order diagrams for double Higgs production from pp collisions. The diagram to the right shows the possibility of a heavy Higgs decaying into a pair of lighter SM Higgs bosons.

Journal of Physics: Conference Series **645** (2015) 012017 doi:10.1088/1742-6596/645/1/012017

References

[1] Aad G *et al* [ATLAS Collaboration] 2012 *Phys. Lett. B* **716** 1 (*Preprint* 1207.7214)
[2] Chatrchyan S *et al* [CMS Collaboration] 2012 *Phys. Lett. B* **716** 30 (*Preprint* 1207.7235)
[3] ATLAS Collaboration 2014 *JHEP* **1409** 112 (*Preprint* 1407.4222)
[4] ATLAS Collaboration 2014 *Phys. Lett. B* **738** 234–253 (*Preprint* 1408.3226)
[5] Chatrchyan S *et al* [CMS Collaboration] 2014 *Phys. Rev. D* **89** 092007 (*Preprint* 1312.5353)
[6] Ade PAR *et al* [Planck Collaboration] Placnk 2013 results. XVI. Cosmological parameters (*Preprint* 1303.5076)
[7] Basso L *et al* 2011 *JHEP* **1108** 122 (*Preprint* 1106.4762)
[8] Alwall J, Herquet M, Maltoni F, Mattelaer O and Stelzer T 2011 *JHEP* **1106** 128 (*Preprint* 1106.0522)
[9] Alves A, Berlin A, Profumo S and Queiroz F S Dark matter complementarity and the z' portal (*Preprint* 1501.03490)
[10] Deshpande N and Ma E 1978 *Phys. Rev. D* **18** 7
[11] ATLAS Collaboration 2015 *Phys. Rev. Lett.* **114** 081802 (*Preprint* 1406.5053)

Journal of Physics: Conference Series **645** (2015) 012018 doi:10.1088/1742-6596/645/1/012018

Towards Optimal Filtering on ARM for ATLAS Tile Calorimeter Front-End Processing

Mitchell A. Cox

School of Physics, University of the Witwatersrand, Johannesburg 2050, South Africa

E-mail: `mitchell.cox@students.wits.ac.za`

Abstract. The Large Hadron Collider at CERN generates enormous amounts of raw data which presents a serious computing challenge. After planned upgrades in 2022, the data output from the ATLAS Tile Calorimeter will increase by 200 times to over 40 Tb/s. Advanced and characteristically expensive Digital Signal Processors (DSPs) and Field Programmable Gate Arrays (FPGAs) are currently used to process this quantity of data. It is proposed that a cost-effective, high data throughput Processing Unit (PU) can be developed by using several ARM System on Chips in a cluster configuration to allow aggregated processing performance and data throughput while maintaining minimal software design difficulty for the end-user. ARM is a cost effective and energy efficient alternative CPU architecture to the long established x86 architecture. This PU could be used for a variety of high-level algorithms on the high data throughput raw data. An Optimal Filtering algorithm has been implemented in C++ and several ARM platforms have been tested. Optimal Filtering is currently used in the ATLAS Tile Calorimeter front-end for basic energy reconstruction and is currently implemented on DSPs.

1. Introduction

Projects such as the Large Hadron Collider (LHC) generate enormous amounts of raw data which presents a serious computing challenge. After planned Phase-II upgrades in 2022, the raw data output from the ATLAS Hadronic Tile Calorimeter (TileCal) will increase by 200 times to over 40 Tb/s (Terabits/s) [1, 2]. It is infeasible to store this data for offline computation.

A paradigm shift is necessary to deal with these future workloads and the cost, energy efficiency, processing performance and I/O throughput of the computing system to achieve this task are vitally important to the success of future big science projects [3].

ARM System on Chips (SoCs) are found in almost all mobile devices due to their low energy consumption, high performance and low cost [4]. The author is developing an ARM-based PU for the ATLAS TileCal as a high throughput, general purpose co-processor to the read-out system Super Read Out Driver (sROD) which can be used to combat the issue of out of time pile-up. Currently Optimal Filtering is used for signal processing of the raw data, but more sophisticated algorithms may be required in future [5, 7]. A general purpose co-processor is able to run more sophisticated and memory intensive algorithms than an FPGA-based device, although the latency cannot be controlled like it can with an FPGA, which is why the sROD is used in the data path [6].

A brief background on the ATLAS TileCal read out system is provided in Section 2. The results of a C++ Optimal Filtering algorithm on ARM are presented in Section 3 and the

Journal of Physics: Conference Series **645** (2015) 012018 doi:10.1088/1742-6596/645/1/012018

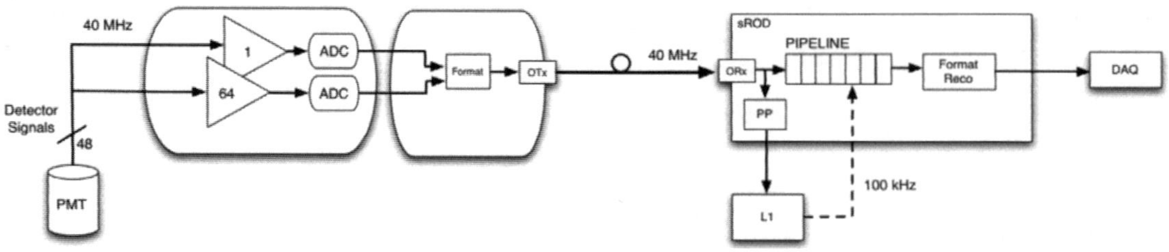

Figure 1: ATLAS TileCal upgraded read out architecture showing the sROD [2].

implementation of a simple PCI-Express CPU interconnect and preliminary performance is given in Section 4. Section 5 concludes with a brief discussion of future work.

2. The ATLAS Tile Calorimeter

The Hadronic Tile Calorimeter (TileCal) is made up by layers of scintillator and steel absorber. When a particle interacts with the scintillator, a pulse of light is produced which is converted to an electrical signal by PMTs. Bunch crossings in the ATLAS detector are separated by 50 ns, with 25 ns planned in future. In reality there can be over 20 separate collisions in a single bunch crossing which leads to a large number of particles interacting with TileCal which is called Pile-Up.

This electrical signal produced by each PMT is conditioned and spread into a pulse with a total length of 150 ns and full width at half maximum of 50 ns. An Analog to Digital Converter (ADC) samples this pulse every 25 ns resulting in seven samples per pulse. A reference pulse showing example sampling points is visible in Fig. 2 with three main parameters of the pulse also illustrated.

The TileCal read out architecture is required to digitize the analog signals produced by the PMTs located on the Tile Calorimeter. In the upgraded system, shown in Fig. 1, all of the PMT data is digitized and sent to the back-end where buffers and processing is done using the sROD [2].

The sROD is located in the back-end, off the detector to avoid the requirement for expensive, radiation-hard electronics. The sROD will be located in an industry standard AdvancedTCA (ATCA) chassis which enables comprehensive redundancy and monitoring to ensure maximum uptime.

In both the existing and the upgraded systems, a pipeline buffer is used to store events until the level one trigger provides an accept signal. This short (typically 5 μs) delay is required while the level one trigger performs computations. In the upgraded system the sROD is able to perform some calculations before sending data to the rest of the triggering and data acquisition system.

Optimal Filtering (OF) or a Matched Filter (MF) are two methods by which the amplitude, A, phase, τ, and base-line pedestal, p, parameters can be calculated from the seven ADC samples of a pulse [7]. The pulse shape and therefore the energy can be reconstructed, when required, from these three parameters [5]. Figure 2 shows an ideal pulse shape with parameters illustrated.

For both the OF and MF algorithms, each parameter (A, τ or p) can be found by multiplying the ADC samples by a specific set of weights which are calculated ahead of time. Both algorithms work well in low luminosity operation where the background noise of the PMT signals is gaussian and uncorrelated. This assumption fails for high luminosity operation (above about $\sqrt{s} = 8$ TeV) where the background noise is no longer uncorrelated due to pile-up [7].

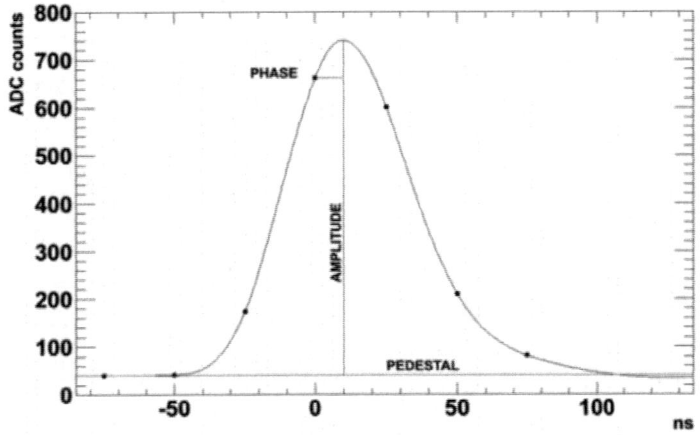

Figure 2: TileCal reference pulse from the PMT showing 7 sample points.

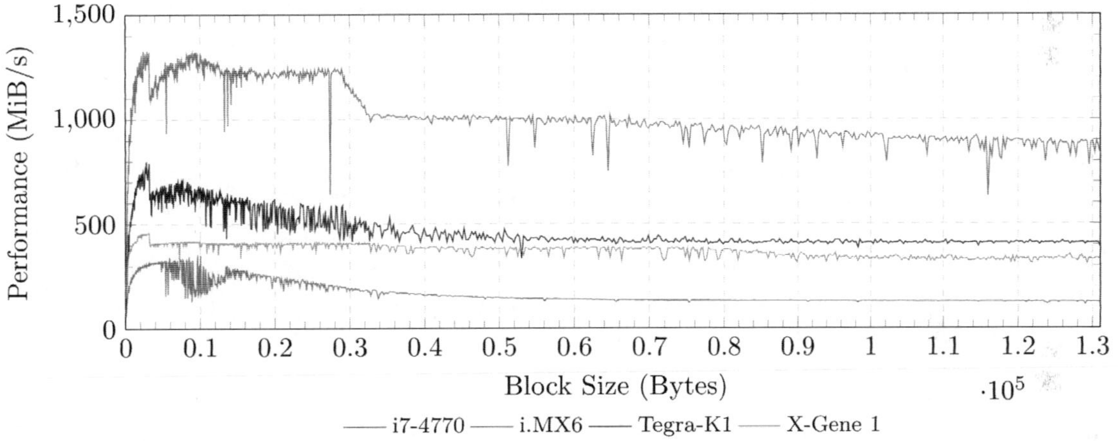

Figure 3: Single thread performance of an Optimal Filtering implementation with seven samples and two outputs with varying block size. An Intel i7-4770 CPU and three ARM platforms are shown.

3. Optimal Filtering on ARM

Optimal Filtering is a simple digital signal processing algorithm where several samples from an analog to digital converter are multiplied with a pre-calculated weight vector. The result of this dot product depends on the set of weights and is tuned to a specific characteristic of the input signal. In the case of the TileCal OF, the peak amplitude and the phase shift (A and τ) of the signal are calculated using two sets of weights.

This algorithm was implemented using the EIGEN C++ library which is one of the fastest linear algebra libraries that also supports ARM [8]. Vector sizes of seven were used to correspond to the current TileCal read-out and two dot products were performed with random weights to represent the amplitude and phase filters.

Figure 3 shows the performance which has been converted to MegaBytes per second (MiB/s) of filtered data where one megabyte is equal to 1024×1024 bytes. A high-end Intel i7-4770 CPU has been compared to an ARM Cortex-A9 (Freescale i.MX6), an ARM Cortex-A15 (NVIDIA Tegra-K1) and an ARMv8 (APM X-Gene 1) platform. The ARMv8 is the only currently

Table 1: PCI-Express throughput results of an i.MX6 pair.

	CPU memcpy	**DMA (EP)**	**DMA (RC)**
Read (MB/s)	$94.8 \pm 1.1\%$	$174.1 \pm 0.3\%$	$236.4 \pm 0.2\%$
Write (MB/s)	$283.3 \pm 0.3\%$	$352.2 \pm 0.3\%$	$357.9 \pm 0.4\%$

available 64 bit ARM System on Chip and is similar to a Cortex-A57.

Batches of seven samples were combined into blocks for more efficient processing. Block sizes from 28 B (one set of samples) to 128 kB were tested.

It is clear that the Intel performance is better than the ARM platforms. It is interesting to note that the performance of the Tegra-K1 is better than the X-Gene. This should not be the case but it was found that the compiler is not yet mature for the new ARMv8 architecture. Using GCC 4.9 instead of GCC 4.8 makes approximately 10% performance difference on the X-Gene but for uniformity GCC 4.8.2 was used on all platforms.

The Wandboard is capable of a peak of 350 MiB/s and the other platforms all achieve over 1 GiB/s of Optimal Filtering throughput. This is typically more than the external I/O available on the SoCs. It is important to note that the results are for a single thread and so if multiple cores are used then the results should scale accordingly.

4. PCI-Express CPU - CPU Interconnect

The external I/O interface of a computing system should be well balanced with the available processing power. Because this is highly dependent on the algorithms used, benchmarking is important. Based on the results in Section 3, even the lowest performing system, the Wandboard, requires up to 350 MiB/s external I/O for good system balance. Gigabit Ethernet, which is almost always available on a SoC, only allows up to 125 MB/s (119 MiB/s). This is clearly insufficient for optimal system balance in this application.

PCI-Express is a high-bandwidth external I/O interface that is energy efficient and simple to implement for the system designer as it only requires several PCB traces and no special and potentially expensive hardware. There is still a requirement to eventually use a standard interconnect such as Ethernet if connecting to an existing system, but this could be accomplished using a single fast adapter node such an a more expensive SoC or some other dedicated system. For the connection to the sROD, PCIe can be used directly.

PCI-Express throughput tests have been performed on a pair of i.MX6 quad-core ARM Cortex-A9 SoCs clocked at 1 GHz, located on Wandboard development boards. The results are presented in Table 1. Three tests were run to ascertain the maximum data throughput that can be obtained from the i.MX6 SoC: a simple CPU based memcpy command and two Imaage Processing Unit based Direct Memory Access (DMA) transfers, initiated by the Endpoint (EP) or slave and the Root Complex (RC) which is the host. The i.MX6 has no dedicated DMA controller on it's PCI-Express controller.

The theoretical maximum throughput for the PCI-Express Gen 2 x1 link used is 500 MB/s. The best result is using DMA initiated by the RC but it is only 72% of the theoretical maximum. The RC-mode drivers are more optimized than the EP-mode drivers due to limited manufacturer support for EP-mode. The read results are lower than write because of overheads to initiate the read.

It is not trivial to use the IPU DMA for generic data transfer as the data is reformatted during movement. This is typical for image based data (scaling, pixel format, etc.) but would increase overhead in the end application that uses the data.

The X-Gene and Tegra-K1 system on chips have more advanced PCI-Express controllers which do support DMA. A test system for these platforms has not been built but in theory there should be no significant issues.

5. Discussion, Conclusions and Future Work

High data throughput computing is required for projects such as the LHC which produce enormous amounts of raw data. A general purpose ARM System on Chip-based processing unit is being developed which will be used as a co-processor to the sROD to help mitigate the energy reconstruction issues caused by pile-up under higher luminosity operation of the LHC.

A PCI-Express interface will be used for the raw data transfer between the sROD and the PU. 2.8 GiB/s of data throughput is required to sustain the raw data from the sROD prototype. Initial throughput measurements presented for a pair of Freescale i.MX6 quad-core Cortex-A9 SoCs are between 283 and 350 MiB/s of the theoretical maximum 488 MiB/s (500 MB/s) for the available x1 link.

An Optimal Filtering algorithm was implemented and tested on ARM Cortex-A9, A15 and X-Gene (similar to Cortex-A57) SoCs. The slowest SoC, the Cortex-A9, achieved 350 MiB/s throughput on the algorithm. The use of gigabit Ethernet would not lead to an optimal system balance between CPU and external I/O and PCIe is the only alternative in this case. On the other platforms the system balance by using gigabit Ethernet is even worse.

The X-Gene SoC supports 10 Gb/s Ethernet which is approximately 1 GB/s of external I/O. This would result in a good system balance if only one CPU core was used. Four or eight cores are present on the X-Gene and so PCI-Express, where the maximum throughput is almost 16 GB/s on the X-Gene, would be a very powerful and flexible solution as an sROD co-processor or general purpose processing unit.

Acknowledgements

The financial assistance of the National Research Foundation towards this research is hereby acknowledged. We would also like to acknowledge the School of Physics, the Faculty of Science and the Research Office at the University of the Witwatersrand, Johannesburg.

References

[1] The ATLAS Collaboration 2010 *The European Physical Journal C* **70** 1193–1236
[2] Carrió F, Castillo V, Ferrer A, Fiorini L, Hernández Y, Higón E, Mellado B, March L, Moreno P, Reed R, Solans C, Valero A and Valls J A 2014 *Journal of Instrumentation* **9** C02019–C02019
[3] Rajovic N, Vilanova L, Villavieja C, Puzovic N and Ramirez A 2013 *Journal of Computational Science* **4** 439–443
[4] Cox M A and Reed R 2014 *Proceedings of SAIP2014, the 59th Annual Conference of the South African Institute of Physics* ed Engelbrecht C and Karataglidis S (Johannesburg) pp 180–184 ISBN 978-0-620-65391-6 URL `http://events.saip.org.za`
[5] Fullana E, Castelo J, Castillo V, Cuenca C, Ferrer A, Higon E, Iglesias C, Munar A, Poveda J, Ruiz-Martinez A, Salvachua B, Solans C and Valls J 2005 Optimal Filtering in the ATLAS Hadronic Tile Calorimeter URL `http://ific.uv.es/tical/Publications/InternalNotes/tilecal-2005-001.pdf`
[6] Cox M A, Reed R and Mellado B 2015 *Journal of Instrumentation* **10** C01007–C01007
[7] Peralva B S 2013 *Nuclear Science Symposium and Medical Imaging Conference (NSS/MIC)* (Seoul: IEEE) pp 1–6 ISBN 978-1-4799-0533-1
[8] Jacob B and Guennebaud G 2014 Eigen C++ Library URL `http://eigen.tuxfamily.org`

Journal of Physics: Conference Series **645** (2015) 012019 doi:10.1088/1742-6596/645/1/012019

Radiation hardness of plastic scintillators for the Tile Calorimeter of the ATLAS detector

H Jivan[1,2], E Sideras-Haddad[1,2], R Erasmus[1], S Liao[1], M Madhuku[3], G Peters[1,2], K Sekonya[1,3], O Solvyanov[4]

1: School of Physics, University of the Witwatersrand, Johannesburg 2050, South Africa
2: DST-NRF Centre for Excellence in Strong Materials
3: iThemba LABS, North, Empire Road, Braamfontein 2000, Johannesburg.
4: European Organization for Nuclear Research (CERN), CH-1211 Genève 23, Suisse

Email: harshna.jivan@gmail.com

Abstract The radiation damage in polyvinyl toluene based plastic scintillator EJ200 obtained from ELJEN technology was investigated. This forms part of a comparative study conducted to aid in the upgrade of the Tile Calorimeter of the ATLAS detector during which the Gap scintillators will be replaced. Samples subjected to 6 MeV proton irradiation using the tandem accelerator of iThemba LABS, were irradiated with doses of approximately 0.8 MGy, 8 MGy, 25 MGy and 80 MGy. The optical properties were investigated using transmission spectroscopy and light yield analysis whilst structural damage was assessed using Raman spectroscopy. Findings indicate that for the dose of 0.8 MGy, no structural damage occurs and light loss can be attributed to a breakdown in the light transfer between base and fluor dopants. For doses of 8 MGy to 80 MGy, structural damage leads to possible hydrogen loss in the benzene ring of the PVT base which forms free radicals. This results in an additional absorptive component causing increased transmission loss and light yield loss with increasing dose.

1. Introduction

The Tile Calorimeter of the ATLAS detector, is a hadronic calorimeter responsible for recording the energy and trajectory of hadrons, taus as well as jets of quarks and gluons that result from the proton-proton collisions within the Large Hadron Collider of CERN. Plastic scintillators form an integral component [1] of this calorimeter due to their ability to undergo fluorescence whereby incident radiation causes electronic excitations within the scintillator material which then relax back to their ground state through the emission of light [2]. Plastic scintillators in specific are chosen for their properties of high optical transmission and fast rise and decay time. This enables efficient data capture since fast signal pulses can be generated [2]. In addition, plastic scintillators are easier to manufacture as compared to inorganic crystals which require special growing methods, and are more cost effective [3] for covering the typically large detector areas. The main problem encountered by plastic scintillators however, is radiation damage incurred due to their interaction with the ionizing particles to be detected. This damage causes a significant decrease in the light yield of the scintillator and introduces an error into the time-of flight data acquired.

As part of the phase two upgrade of the Tile Calorimeter planned for 2018, a comparative study is underway which examines the radiation hardness of several plastic scintillators. The results will aid in

Journal of Physics: Conference Series **645** (2015) 012019 doi:10.1088/1742-6596/645/1/012019

choosing the best grade for replacing the current crack scintillators within the Gap region of the Tile Calorimeter as well as provide insight into the type of damage that occurs. In these proceedings, we present preliminary results for EJ200, a polyvinyl toluene based plastic obtained from ELJEN Technologies.

2. Scintillator material under study

Plastic scintillator EJ200 is composed of a polyvinyl toluene base and 3% of added organic fluors. The absorption peak of the fluors coincide with the emission peak of the base. This allows for the fluors to absorb light emitted by the base and re-emit it at higher wavelengths, thus preventing re-absorption within the base. As a result, the scintillator becomes more transparent to its own light and the light output is boosted. Some important properties of EJ200 are listed in table 1 below.

Table 1: Properties of plastic scintillator EJ200 as provided by the manufacturers [4]

	EJ200
Light Output (% Anthracene)	64
Wavelength of maximum emission (nm)	425
Rise time (ns)	0.9
Decay time (ns)	2.1
Refractive index	1.58

3. Experimental Procedure

The tandem accelerator of iThemba LABS in Gauteng was used to irradiate samples with 6 MeV protons. To ensure that the study simulated a similar type of particle-scintillator interaction as observed in the Tile Calorimeter, protons would be required to pass through the samples whilst imparting energy primarily through ionisation losses. In this manner, only electronic stopping is taking place while nuclear stopping which is responsible for dislocations is prevented. In order to ensure this, SRIM simulations were performed to determine the stopping range of 6 MeV protons within the plastic material. Based on these simulations, several samples were cut and polished to dimensions of 0.5 cm by 0.5 cm, with thickness of ~350 μm. A polishing procedure based on standard metallographic techniques was employed. Samples were irradiated to doses of 0.8 MGy, 8 MGy, 25 MGy and 80 MGy. A set of magnetic quadrupole triplets was used to focus the proton beam to a spot size of ~20-30 μm and the beam was then scanned in the x and y plane to achieve an irradiated area of approximately 1.8 mm by 1.8 mm.

The effects of proton damage on the optical properties of the samples were characterised by conducting light transmission and light yield studies. Transmission spectroscopy was conducted using the Varian Carry 500 spectrophotometer. Light transmission was measured relative to transmission in air over a range of 200-800 nm. Transmission spectra were collected prior to irradiation as well as over several days after. The scintillator light yield in response to 0.5 MeV gamma rays emitted by a ^{90}Sr source was measured using the light box set-up at CERN. In these experiments, the sample was coupled to two Kuraray Y-11(200) optical fibers which were connected to a standard Tile Calorimeter photomultiplier tube. Light yield experiments were conducted several weeks after irradiation thus providing time for partial recovery in the scintillators.

The structural damage undergone by the scintillators were then investigated using Raman spectroscopy. Raman spectra were obtained for the un-irradiated control samples as well as on the 0.8 MGy and 8 MGy irradiated samples using the Horiba Jobin-Yvon Raman spectrograph. An Argon laser was used to provide a 515 nm excitation wavelength. Raman studies were conducted 3 days, 10 days and 4 weeks after irradiation.

4. Light Transmission Results and Analysis

The ratio of transmission in each irradiated EJ200 sample relative to its transmission before irradiation for several doses as a function of wavelength is shown in Fig. 1 below. Absorption in the PVT base occurs over 240 nm to 300 nm whilst absorption in the fluors is expected to occur over 270 nm to 400 nm. For the dose of 0.8 MGy, a transmission dip at lower wavelengths is observed, followed by a peak over 320-400 nm. This feature indicates that a competitive absorptive process has developed between the primary and secondary fluors.

As the dose increases, we observe the formation of an absorptive tint whereby the absorption drop-off shifts to higher wavelengths. This implies that additional absorbing species such as free radicals are forming within the scintillator and leads to a further decrease in light emitted by the scintillator. This additional absorption component may also reduce the attenuation length of the scintillation light as described by C. Zorn (1993) [5].

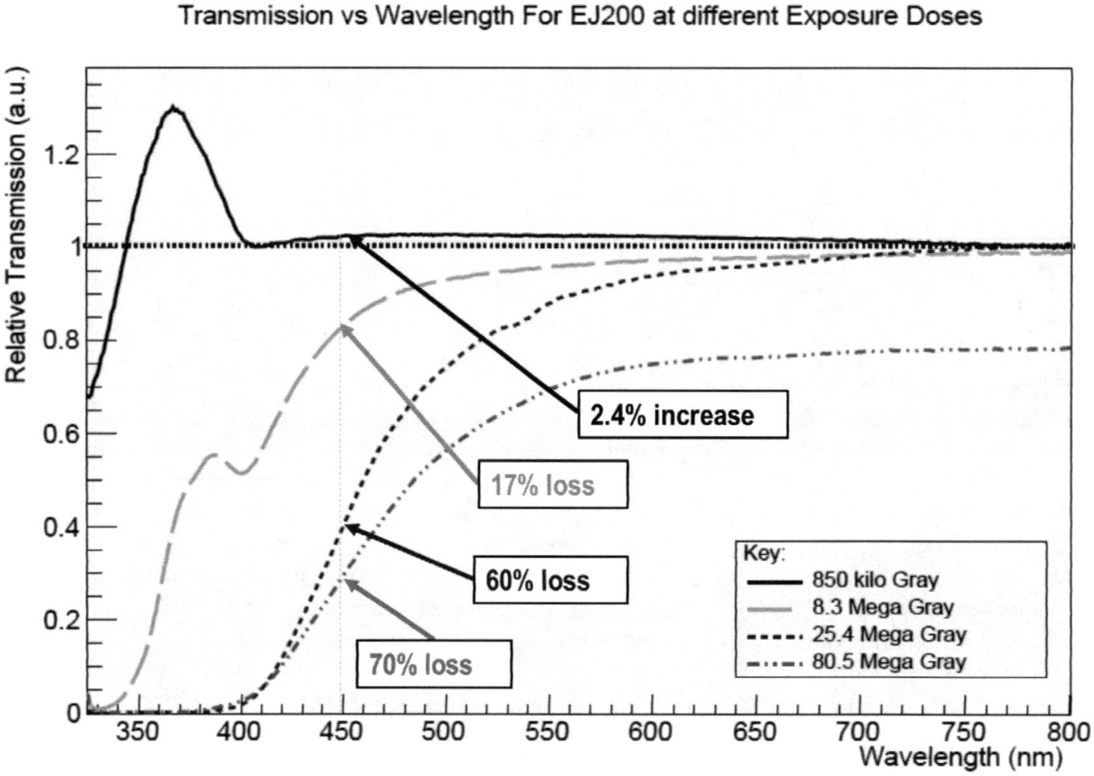

Figure 1: Ratio of transmission in irradiated sample to transmission before irradiation for EJ200 samples irradiated over several doses.

5. Light Yield Results and Analysis

Experiments conducted on the light output from the scintillator samples responding to 0.5 MeV gamma rays emitted from ^{90}Sr were repeated thrice per sample and averaged. The light response in the irradiated samples as a ratio to light output in an un-irradiated control sample as a function of irradiation dose is shown in Figure 2. There appears to be a linear decreasing trend in light loss with increasing orders of magnitude in dose. However, these results are very preliminary as geometric considerations in the experimental set-up need to be accounted for. This may account for the large variation in light loss seen between two samples irradiated to a similar dose of ~ 0.9-1 MGy.

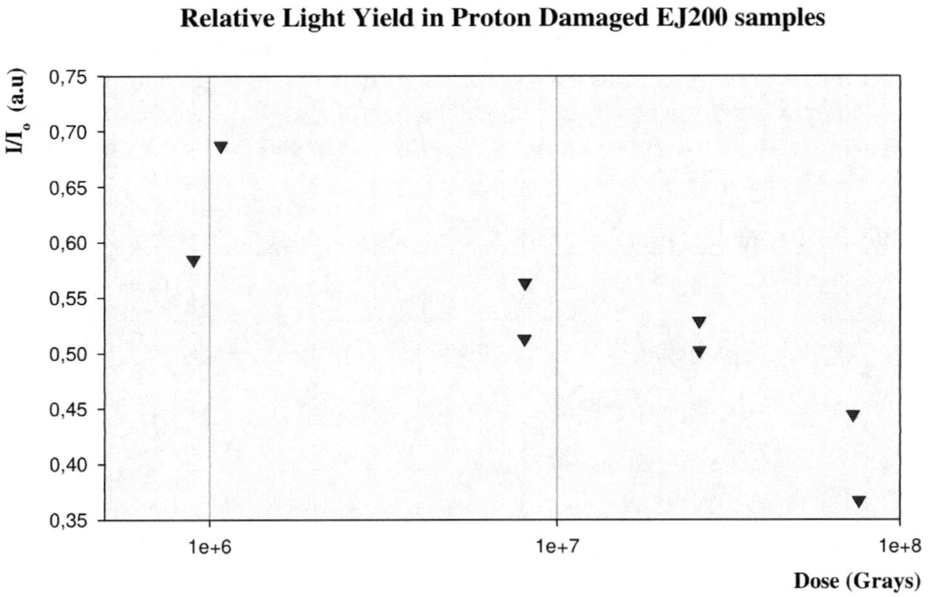

Figure 2: Relative light yield in EJ200 samples as a function of irradiation dose.

6. Raman Spectroscopy Results and Analysis

Raman spectra were obtained for the 0.8 MGy samples, 8 MGy samples as well as for un-irradiated control samples. The samples irradiated to 0.8 MGy maintained their structure and very little damage was observed. In the 8 MGy sample, an increased amount of background fluorescence could be observed which may be due to free radicals interacting with the excitation light. Raman spectroscopy could not be performed for the high dose of 80 MGy because this background fluorescence becomes too large to obtain an experimentally viable spectrum. Figure 3 shows the background subtracted Raman spectrum for the 8 MGy irradiated sample taken over several days.

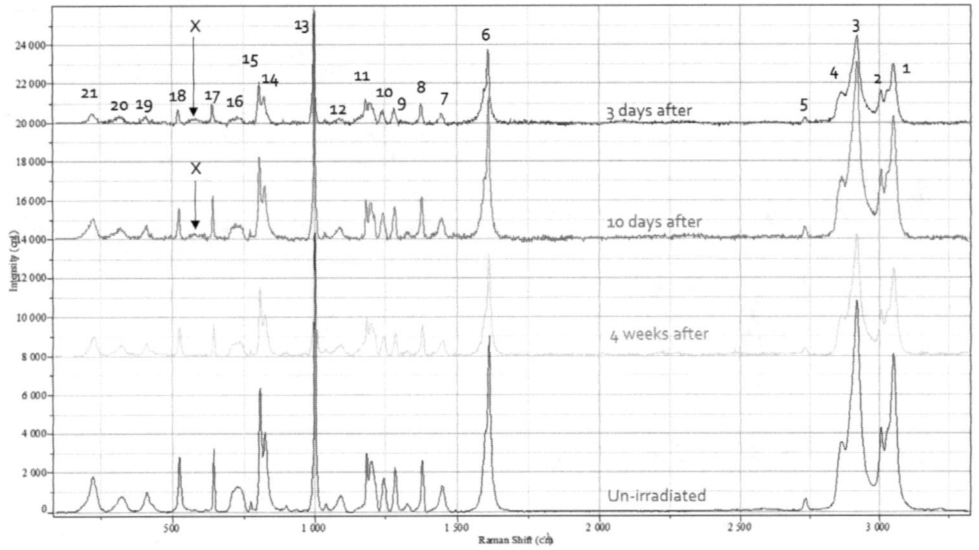

Figure 3: Background subtracted Raman spectra for 8 MGy irradiated EJ200.
N.B Spectra overlap but intensities have been shifted in order to have better visible representation.

Journal of Physics: Conference Series **645** (2015) 012019 doi:10.1088/1742-6596/645/1/012019

The peaks observed are allocated to the characteristic functional groups or vibrational groups using the Raman Peak assignment datasheet from Horiba Jobin-Yvon. The peak assignment is summarized in Table 2. An additional peak (X) was found to form after irradiation but disappeared over time as the scintillator healed. In order to gauge changes in the amount of specific species present, the intensities of the various peaks were plotted relative to peak 13 which represents the C-C bond typically found in benzene rings. These results are shown in Figure 4.

Table 2: Allocation of peaks to characteristic functional groups or vibrational groups.

Functional Group/ Vibration	Peak Assignment
δ(C-C) aliphatic	20-21
ν(C-C) alicyclic or aliphatic chain vibrations	9-12, 14-19, X
ν(C-C) aromatic ring chain vibrations	13
δ(CH3)	8
δ(CH2) or δ(CH3)asymmetric	7
ν(C=C)	6
ν(C-H)	3-4
ν(=(C-H))	1-2

The ratio of species typically found in the vinyl backbone of the sample versus that in the benzene ring shows an increase after irradiation. A small amount of additional alicyclic rings and aliphatic chains are found with the additional peak X being typical of these bonds. These bonds represent single carbon rings and chains stripped of hydrogen. This indicates that structural damage occurs within the scintillator whereby bonds in the benzene ring are broken. Hydrogen may be lost by the benzene rings to form free radicals. After 4 weeks of healing, the scintillator appears to substantially recover its structure.

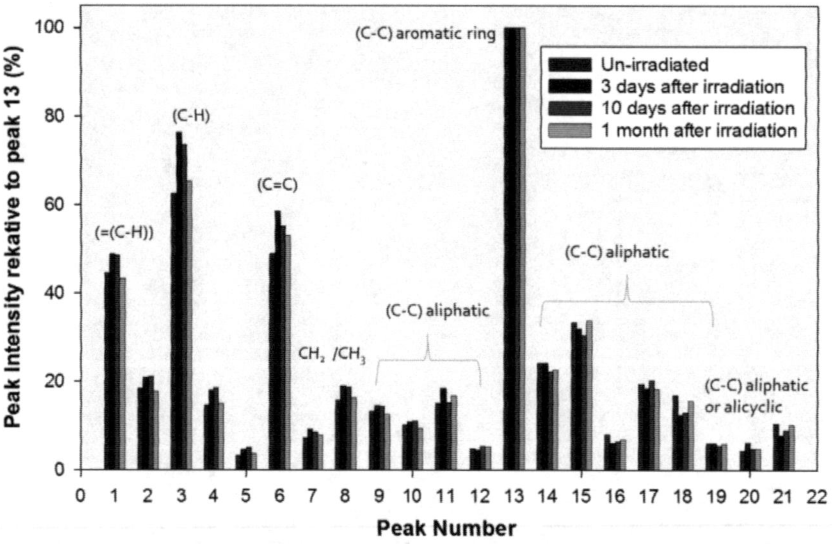

Figure 4: Plot of the intensities of peaks relative to peak 13 representing the C-C aromatic ring bond.

Journal of Physics: Conference Series **645** (2015) 012019 doi:10.1088/1742-6596/645/1/012019

7. Conclusion

According to the results obtained in this study, irradiation to high doses in plastic scintillator EJ200 using 6 MeV protons causes radiation damage and leads to a reduced performance in its scintillation capabilities. At a dose of approximately 0.8 MGy, the structure of the scintillator remains robust, however light loss is still observed. This can be accounted for by noting that light transmission indicates a competitive absorptive process developing between the primary and secondary fluors.

At a dose of 8 MGy, more light loss is observed and the scintillator becomes less transparent to its own light. The formation of an absorptive tint indicates that free radicals may be absorbing light that would otherwise be collected from the scintillator. These free radicals may arise from hydrogen that is lost through bonds breaking within the benzene rings of the polyvinyl toluene base and hence leading to additional aliphatic chains and alicyclic rings observed in the Raman spectra. The Raman spectra further indicate that after 4 weeks, the scintillator significantly recovers its structure. As the irradiation dose is increased, a greater loss to light transparency is observed. This can also be visibly noticed in the increased yellowing of the irradiation spot among samples with progressing higher doses. Increasing dose leads to greater structural damage and leads to a reduction in the light output of the scintillator.

Overall, EJ200 can be considered as substantially radiation hard since for a dose of 0.8 MGy, it still maintains its structural composition, remains fairly transparent and its light output is still greater than 50% of its original capability. It is only at doses higher than 8 MGy that permanent structural damage occurs. The highest dose currently expected within the Tile Calorimeter over the next several years is 10 kGy and therefore EJ200 may be a potential candidate for consideration during the 2018 replacement of the GAP scintillators.

8. Upcoming work

This study is being carried out on several other plastic scintillators including PVT based EJ208, EJ260 and BC-408 obtained from ELJEN technologies and Saint Gobain Crystals respectively. These commercial scintillators will be compared to the polystyrene based scintillators presently used within the Tile Calorimeter. In addition to the techniques looked at in this paper, healing in the scintillators as well as the effect of dose rate on the damage induced are being looked at. Electron Paramagnetic resonance (EPR) and nuclear magnetic resonance (NMR) studies are also being conducted in order identify the types of free radicals formed due to radiation damage.

9. References

[1] ATLAS Collaboration 2008 *The ATLAS Experiment at the CERN Large Hadron Collider* (IOP Publishing and SISSA)

[2] Knoll G F 1999 *Radiation Detection and Measurement, Third Edition* (Michigan: John Wiley & Sons Inc.) chapter 8 pp 220-222

[3] Chen M 2011 *Queen's University, PHYS 352: Measurement, Instrumentation and Experiment Design* [Online]. Available: http://www.physics.queensu.ca/~phys352/lect19.pdf.

[4] ELJEN Technology 2013 *Products: Plastic Scintilators* [Online]. Available: http://www.eljentechnology.com/index.php/products/plastic-scintillators.

[5] Zorn C 1993 A pedestrian's guide to radiation damage in plastic scintillators *Radiat. Phys. Chem.* **41** 37-43

Design of a new front-end electronics test-bench for the upgraded ATLAS detector's Tile Calorimeter

C O Kureba, M Govender, I Hofsajer, X Ruan, C Sandrock, M Spoor

School of Physics, University of the Witwatersrand, Johannesburg 2050, South Africa

E-mail: Chamunorwa.Kureba@wits.ac.za

Abstract. The year 2022 has been scheduled to see an upgrade of the Large Hadron Collider (LHC), in order to increase its instantaneous luminosity. The High Luminosity LHC, also referred to as the upgrade Phase-II, means an inevitable complete re-design of the read-out electronics in the Tile Calorimeter (TileCal) of the A Toroidal LHC Apparatus (ATLAS) detector. Here, the new read-out architecture is expected to have the front-end electronics transmit fully digitized information of the detector to the back-end electronics system. Fully digitized signals will allow more sophisticated reconstruction algorithms which will contribute to the required improved triggers at high pile-up. In Phase II, the current Mobile Drawer Integrity ChecKing (MobiDICK) test-bench will be replaced by the next generation test-bench for the TileCal superdrawers, the new Prometeo (A Portable ReadOut ModulE for Tilecal ElectrOnics). Prometeo is a portable, high-throughput electronic system for full certification of the front-end electronics of the ATLAS TileCal. It is designed to interface to the fast links and perform a series of tests on the data to assess the certification of the electronics. The Prometeo's prototype is being assembled by the University of the Witwatersrand and installed at CERN for further developing, tuning and tests. This article describes the overall design of the new Prometeo, and how it fits into the TileCal electronics upgrade.

1. Introduction

Scientists and engineers at the European Organization for Nuclear Research (CERN) probe the fundamental structure of matter. Here, the LHC accelerates and collides protons, and also heavy ions. ATLAS [1] is one of two general purpose detectors used for detecting the sub-atomic particles produced during these high-energy collisions. TileCal [2] is the central hadronic calorimeter of the ATLAS detector. Alternating steel plates and plastic scintillator tiles make up each cell of TileCal. The energy of the detected particles is sampled by the scintillators and signals are collected from the photomultipliers (PMTs) by the front-end electronics located in the superdrawers, which is the outermost part of the detector.

In order to certify the TileCal as being ready for data collection, an evaluation must be made on each of its modules, hence, the need for an electronics test-bench. Currently, the MobiDICK4 test-bench is being used for the certification process. The High Luminosity LHC will lead to an inevitable complete re-design of the read-out electronics in the TileCal. Figure 1 shows a comparison of the current and future front-end electronics set-up for the TileCal. In the current set-up, pipeline memories are used to store digitized data samples before they can be trigger-selected. The read-out electronics systems in each superdrawer are daisy-chained, resulting in the sharing of a data connection to the Read-Out Driver (ROD). In the future electronics design,

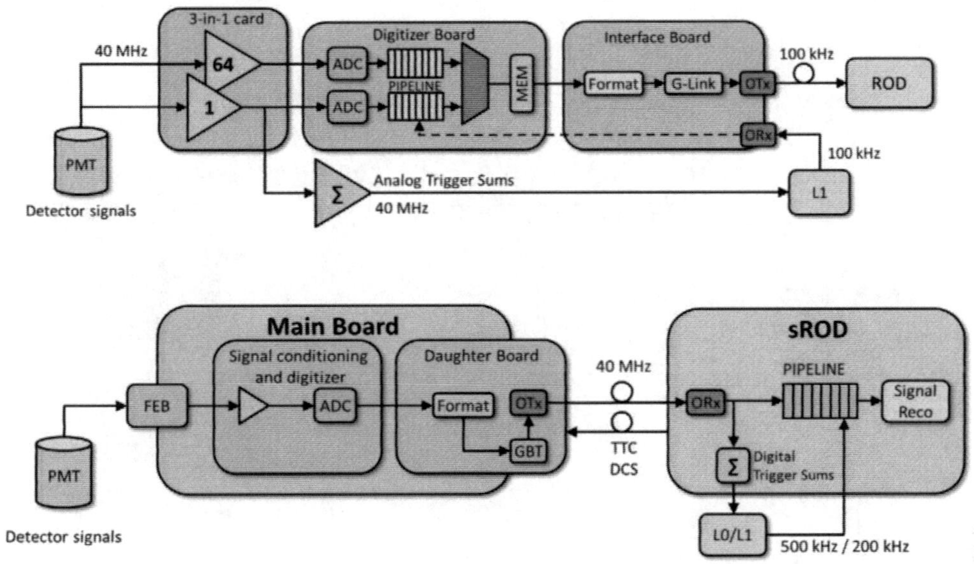

Figure 1. *Top*: Current front-end electronics; *Bottom*: Equivalent electronics for Phase-II upgrade.

each superdrawer has been divided into 4 minidrawers. One minidrawer hosts 12 PMTs and 1 daughter board. Each daughter board has one link to the super Read-Out Driver (sROD), thereby reducing failure rate by 25% with respect to ATLAS. There are up to 48 PMTs in one superdrawer, grouped in groups of 6 in a digitizer board. There are 16 digitizers for one superdrawer, interfaced by 1 interface card. It is envisaged that the future electronics will have:

- a super Read-Out Driver (sROD), capable of receiving data at 40 MHz as opposed to the current Read-Out Driver (ROD) which only handles 100 kHz,
- an increase in the number of point-to-point links with the front-end electronics,
- improved radiation tolerance,
- a higher read-out bandwidth due to the need to read-out all sampled data to avoid corruption in the front-end pipeline memories.

An evaluation of this new proposed architecture is currently being carried out in the demonstrator project, where a small fraction of the detector (1/256) will be evaluated in test beams and inserted into ATLAS at the next shutdown of the LHC. In Phase II, the current Mobile Drawer Integrity ChecKing (MobiDICK4) system [4, 5] test-bench will be replaced by the next generation test-bench for the TileCal superdrawers, the new Prometeo (A Portable Read- Out ModulE for Tilecal ElectrOnics). The MobiDICK system faces challenges against ageing and new technologies [6]. The Prometeo is designed to certificate the TileCal front-end electronics by performing multiple tests. The Prometeo's prototype is being assembled by the University of the Witwatersrand and installed at CERN for further developing, tuning and tests.

2. MobiDICK4 (Current test-bench)

Figure 2 shows a layout of the components of the MobiDICK4 test-bench, currently being used at CERN for the certification of the TileCal front-end electronics. Its main board is a Xilinx ML507 evaluation board, whose backbone is a Virtex-5 Field Programmable Gate Array (FPGA). The MobiDICK4 test-bench has a server running on Power PC, which connects to the client via

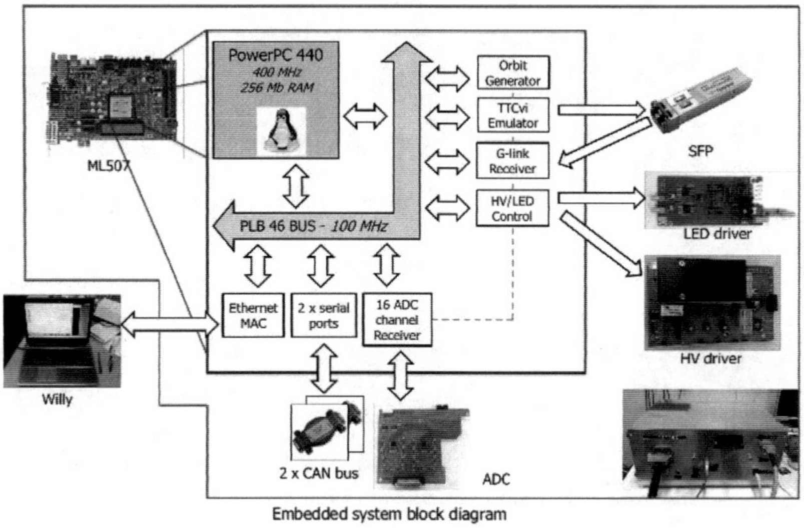

Figure 2. Layout of the MobiDICK4 test-bench, currently being used at CERN.

ethernet. It records digital data at a maximum rate of 100 kHz. It uses "slow" canbus to control the current integrator circuit and the HV applied to every PMT and a custom designed ADC board to sample the analogue signals.

3. Prometeo (new test-bench): Hardware design

Prometeo has been designed to have the ability to: read-out all channels at the LHC bunch crossing frequency, assess the quality of data in real-time, diagnose malfunctions in each minidrawer be self-contained and portable for maintenance inside the detector, and be low-cost and scalable for network usage. In the current design being deployed for the hybrid demonstrator that will also have analogue output, Prometeo will make use of an ADC board. The current version is using a VC707 board. The prototype for 1 superdrawer will use the sROD demonstrator, which is limited to 2 minidrawers. Figure 3 shows a schematic layout of the Prometeo. The Prometeo consists of several parts as follows:

- Main board: Xilinx VC707, Virtex-7 FPGA chip, 1 GB DDR3 RAM, two FMC connectors, received through SFP on the mainboard,
- Dual QSFP FMC card for digital communication with 2 minidrawers,
- Analogue to Digital Converter (ADC) FMC card: for the digitization of data in the hyprid demonstrator,
- High Voltage (HV) board: to turn on/off the HV and provide the -830 V voltage to power on the photo-multipliers (PMTs),
- Light Emitting Diode (LED) board: to illuminate the PMTs,
- Commercial ATX power supply,
- Ethernet router,
- Aluminum enclosure: whose dimensions are 50 cm × 35 cm × 20 cm in terms of length, width and depth, thickness of 3 mm and weight of 8 kg.

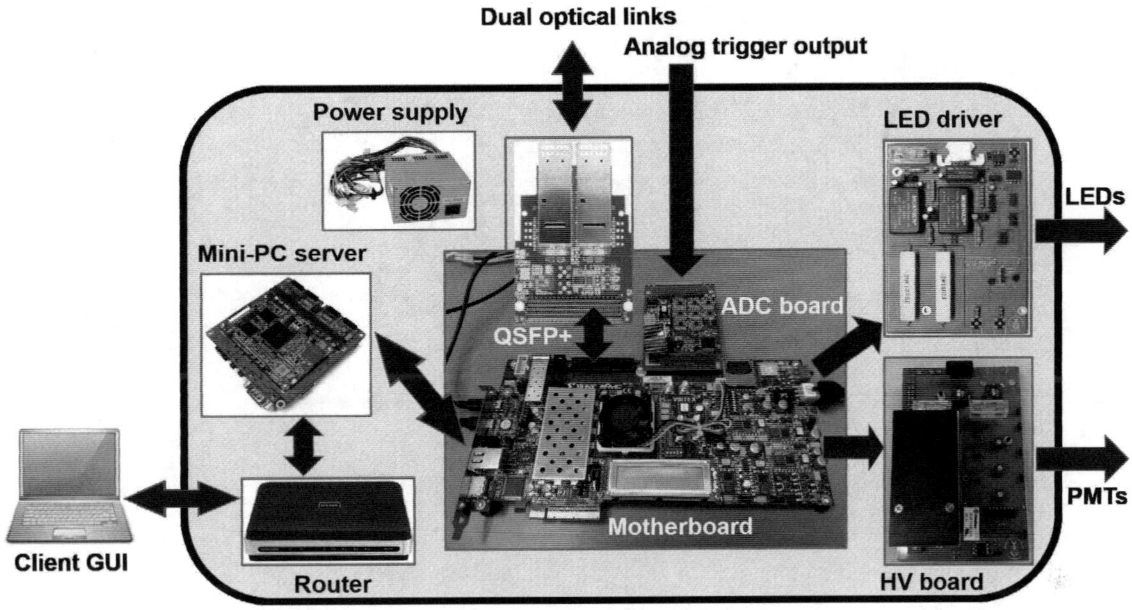

Figure 3. Schematic of the Prometeo.

4. Prometeo: Software design

The software design of Prometeo is based on direct hardware access technology for FPGAs with direct ethernet connection. We use an IPbus [7] ip-core module on the FPGA that enables direct communication of the registers through VOP. A custom built software version of the IPbus client that is implemented in C++, Java and Python is used to allow multiple developments and tests to perform in parallel. At this stage there are two implementations of the software being developed, namely, one in pure Java and another one in C++/Python. The Java software is inspired by the one in the MobiDICK GUI where a core window loads different tests as plugin tabs, and each one executes and displays a different test. The C++ one implements the tests directly as command line applications. A front-end HTML/Python interface is used to execute the tests and assess the results. Figure 4 shows snapshots of the Java and HTML GUIs used by the client to execute the Prometeo tests.

The Prometeo has been designed to perform both linearity and stability tests on each of the following:

- Charge Injection System (CIS)
- High Voltage (HV)
- Integrator
- Pedestal

These results are used for diagnosis of faulty PMTs, 3 in 1 cards, mainboards, daughter boards, optical links and analogue cards. Most of the tests are implemented in the HTML interface, thereby leaving Java as an excellent tool for debugging at expert level.

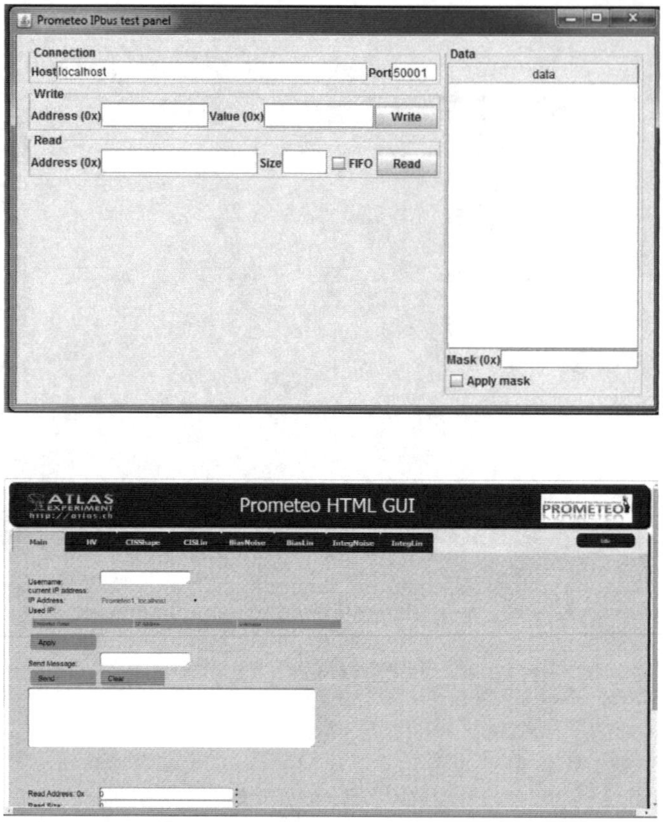

Figure 4. *Top*: Java GUI for Prometeo; *Bottom*: HTML GUI for Prometeo.

5. Conclusion

Prometeo, the next generation test-bench for the certification of the new TileCal electronics has been designed, in preparation for the LHC phase-II upgrade. Prometeo is inspired by the current MobiDICK4 test-bench. It is capable of performing multiple tests on the electronics modules of the upgraded superdrawers. Prometeo is currently in a prototyping phase and currently serves a hybrid demonstrator of the future Phase-II electronics. All components of Prometeo have been manufactured and are undergoing tests as they are added to the system. Further tests are being developed and tested in parallel with the development of the demonstrator. Prometeo software is well advanced and used everyday to assess the stability of the first demonstrator.

References

[1] ATLAS Collaboration The ATLAS Experiment at CERN Large Hadron Collider *JINST* **3** (2008) S08003
[2] ATLAS Collaboration 2010 Readiness of the ATLAS Tile Calorimeter for LHC collisions *Eur. Phys. J. C* **70**
[3] Carrio F *et al* The sROD module for the ATLAS Tile Calorimeter Phase-II upgrade demonstrator *ATL-TILECAL-PROC-2013-020*
[4] Calvet D and Giangiobbe V 2008 Performance of the TileCal superdrawers from a global analysis of the MobiDICK tests *ATL-TILECAL-PUB-2008-007*
[5] Carrio F *et al* Design of an FPGA-based embedded system for the ATLAS Tile Calorimeter front-end electronics test-bench *ATL-TILECAL-PROC-2013-017*
[6] Solans C *et al* 2014 Computing challenges in the certification of ATLAS Tile Calorimeter front-end electronics during maintenance periods *J. Phys.: Conf. Ser.* **513** 012035
[7] Frazier R *et al* The IPbus Protocol: An IP-based control protocol for ATCA/μTCA (2013)*Draft 6*
[8] Moreno P *et al* A new portable test-bench for the ATLAS Tile Calorimeter front-end electronics *JINST* **8** (2013) C02046

Journal of Physics: Conference Series **645** (2015) 012021 doi:10.1088/1742-6596/645/1/012021

A comparative study of the radiation hardness of plastic scintillators for the upgrade of the Tile Calorimeter of the ATLAS detector

S. Liao, R. Erasmus, H. Jivan, C. Pelwan, G. Peters and E. Sideras-Haddad

School of Physics, University of the Witwatersrand, Johannesburg 2050, South Africa

E-mail: `shell-may.liao@cern.ch`

Abstract. The influence of radiation on the light transmittance of plastic scintillators was studied experimentally. The high optical transmittance property of plastic scintillators makes them essential in the effective functioning of the Tile calorimeter of the ATLAS detector at CERN. This significant role played by the scintillators makes this research imperative in the movement towards the upgrade of the tile calorimeter. The radiation damage of polyvinyl toluene (PVT) based plastic scintillators was studied, namely, EJ-200, EJ-208 and EJ-260, all manufactured and provided to us by ELJEN technology. In addition, in order to compare to scintillator brands actually in use at the ATLAS detector currently, two polystyrene (PS) based scintillators and an additional PVT based scintillator were also scrutinized in this study, namely, Dubna, Protvino and Bicron, respectively. All the samples were irradiated using a 6 MeV proton beam at different doses at iThemba LABS Gauteng. The radiation process was planned and mimicked by doing simulations using a SRIM program. In addition, transmission spectra for the irradiated and unirradiated samples of each grade were obtained, observed and analyzed.

1. Introduction

Plastic scintillators are materials that exhibit the property of luminescence when excited by ionizing radiation. This, together with their high optical transmission property makes them suitable for use within large particle detectors such as the ATLAS detector of the Large Hadron Collider (LHC) at CERN [3]. The plastic scintillators under scrutiny are found within the Tile calorimeter of the ATLAS detector. The Tile calorimeter is situated within the hadronic calorimeter of the ATLAS detector, which usually detects hadrons, quarks and gluons. It is 8 metres in diameter and 12 metres in length and makes use of steel as the absorber material, as well as the scintillation plates which are of interest in this research [4]. The steel plates act as an absorption medium and convert the incoming jets into a shower of particles. These showers tend to interact with the plastic tiles and the tiles absorb the energy from these particles and emit the energy in the form of light. The light that is produced within the scintillators is then guided by wavelength shifting fibres and read out through the photomultiplier tubes (PMTs) [4]. The performance of the Tile calorimeter therefore has a direct impact on signatures involving hadrons, jets and missing transverse energy. However, the main problem encountered by these scintillators is radiation damage due to the highly ionizing nature of the particles that are being detected. Thus, due to the anticipated increase in beam energy, there is bound to be an increase in the amount of radiation that the tiles are exposed to. As a result, the need for the use of the most radiation hard plastic scintillators is mandatory in the movement towards upgrading the Tile calorimeter of the ATLAS detector.

2. The scintillation mechanism

The plastic scintillators used in this study were organic Polyvinyl Toluene (PVT) based scintillators doped with fluors. The fast response and low cost production of organic scintillators such as the PVT makes them a rational choice for use in large detectors. They are made up of long chain vinyl toluene molecules. Toluene consists of a benzene ring bonded to a methyl (CH3) and a vinyl group (CH2-CH-) [6]. The benzene ring is a common feature in plastic scintillators and it is characterized by delocalization of 3 π-bonds [1]. The chemical bonds present in a benzene ring include σ-bonds which are in the plane and π-bonds which are out of the plane and overlap. The π-bonds give rise to a cloud of electrons above and below the molecular plane and these π electrons are completely delocalized. Figure 1 below shows the overlap of the π-bonds which give rise to the delocalized electrons.

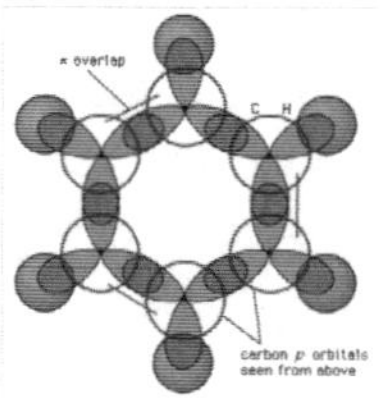

Figure 1. Schematic diagram of the π-bond overlap

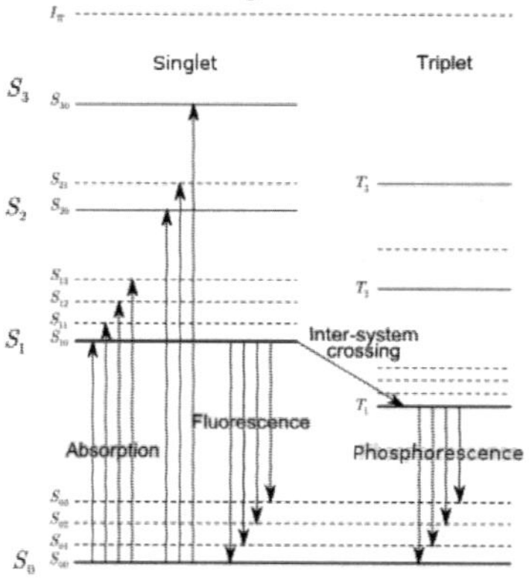

Figure 2. Electronic energy levels of an organic molecule

These delocalized electrons are prone to excitation by radiation, which is essential in the scintillation process. In fact, it is the de-excitation of these delocalized π electrons which result in luminescence. Figure 2 above shows the π-electronic energy levels of an organic molecule. When particles strike the surface of such a scintillator, the radiation is absorbed and results in a molecular excitation to the first excited singlet state. This is almost instantly followed by de-excitation back to the ground state, which is called fluorescence (scintillation). In addition, plastic scintillators are doped with organic fluors: Primary fluor and Secondary fluor. These organic fluors are compounds which absorb the base scintillation and re-emit it at longer wavelengths.

3. Methods

3.1. Modelling the radiation environment

Within the Tile calorimeter, incoming jets carry high energy and interact with the plastic scintillators by imparting energy through this ionization process, and such interactions are essential in understanding the damage mechanism within the target. This calls for a method that one can use to model the radiation environment prior to the actual radiation process in order for effective research results. This modelling was done using SRIM simulations. When the ions interact with the target atoms, they do not get stopped within the material. That is, the ions propagate straight through the material without stopping, hence the need to ensure that the ion beam that will be used during the radiation process also pass straight through the scintillators without stopping. SRIM simulations were then conducted in order to find out whether the proton beam will pass straight through the 350 um thick Polyvinyl Toluene based plastic scintillators, as this was the thickness that we anticipated to cut the samples into. The necessary information that is extracted from the SRIM output is that the 6 MeV proton beam stops at a distance of 474.17 um. Since the samples which are used in this research are 350 um thick, this data tells us that the 6 MeV proton beam will pass straight through the plastic scintillators without stopping. Furthermore, TRIM establishes collisions between the ions and target atoms and thus give information about the energy lost by the protons during ionization.

From this knowledge, we were able to plan the beam time and beam currents needed to achieve the anticipated doses using the following relation:

$$R = \frac{it \times E(lost)}{q \times m} \tag{1}$$

Thus from the simulations, the average energy of the transmitted protons through a 350 um thick PVT target is approximately 2.8MeV, which implies that the energy lost by the 6MeV protons is given by:

$$6MeV - 2.8MeV = 3.2MeV \tag{2}$$

3.2. Sample preparation

Sample preparation was carried out within the physics building Materials Preparation Lab at the University of the Witwatersrand. Initially, the plastic scintillator samples were of dimensions of approximately 200 mm X 25 mm X 10 mm in length, width and thickness, respectively. For the irradiation process the samples needed to be sectioned and cut to dimensions 5 mm X 5 mm X 0.35 mm in length, width and thickness, respectively. Following the cutting, the samples were mounted onto aluminum holders and polished down to the required thickness using a Struers Metallographic polishing machine. After the samples were polished, their thicknesses were measured and documented and it was found that there was a variation in the thicknesses of the samples. That is, they ranged between 300 um and 380 um. Simulations were then done again for this range of thicknesses but in 10 um intervals in order to ensure accurate beam time planning.

4. Experimental procedure

4.1. Radiation process

The radiation of the PVT based plastic scintillators was done using the Tandem accelerator found at iThemba LABS Gauteng. A 6 MeV proton beam was used during the irradiation process. Each brand of PVT based plastic scintillator was exposed to four different doses, a high dose (80 MGy), an intermediate dose (25 MGy), a medium dose (8 MGy) and a low dose (0.8 MGy).

4.2. Transmission spectroscopy

Immediately after irradiation the samples were subject to transmission spectroscopy. That is, data for light transmittance of the PVT based plastic scintillators was taken. This was done using The Varian Cary 500 spectrophotometer which is available within the physics building of the University of the Witwatersrand. Data for transmission data was taken and plotted as percentage transmission versus the wavelength of light. The transmission data obtained was plotted in order to observe and analyze the effects of radiation damage on the light transmittance of the plastic scintillators.

5. Results and analysis

The data was input and plotted using ROOT plot as it is the prominent plotting program used at CERN. In order to gauge whether the decrease in % light transmission increased with the radiation dose, graphs of % transmission versus wavelength for each grade at different doses were plotted, as seen in the case of EJ 208 (Figure 3). Also, in order to observe whether the samples showed some recovery or not following the radiation process, graphs of % transmission versus wavelength for each grade for transmission taken on different days were also plotted, as seen in the case of EJ 208 in Figure 4.

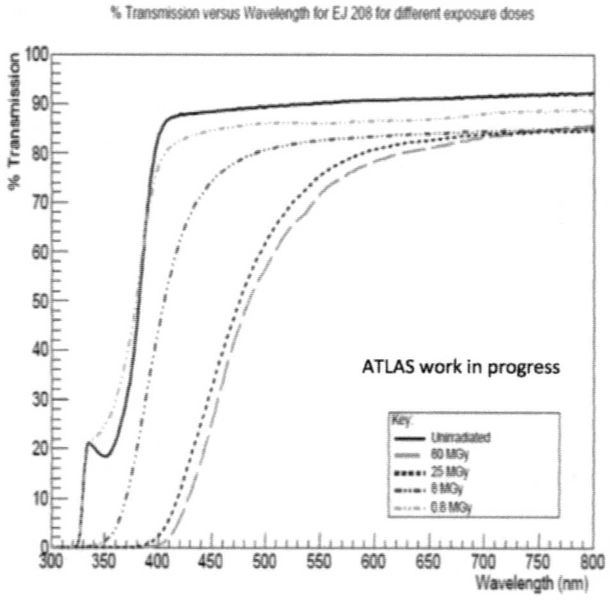

Figure 3. A plot of Transmission versus wavelength for EJ 208 for all doses

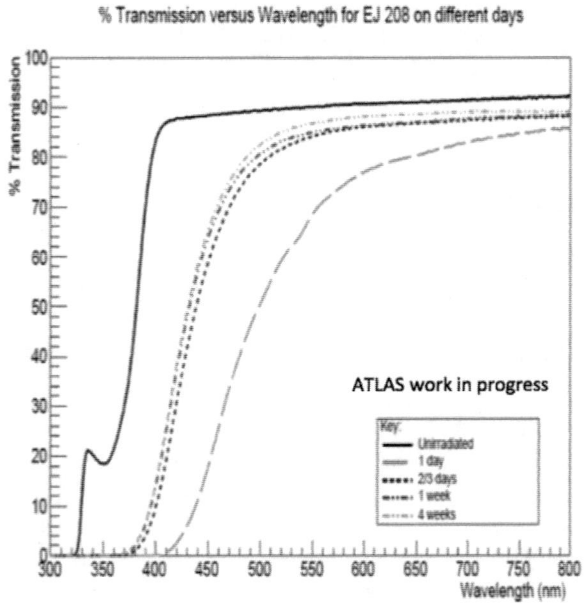

Figure 4. A plot of transmission versus wavelength for EJ 208 on different days

Sample	Dose (MGy)	% Trans. loss	Sample	Dose (MGy)	% Trans. loss
EJ 200	80	42.9	Protvino	80	60.8
	25	28.6		25	34.8
	8	14		8	7.4
	0.8	3.9		0.8	3.3
EJ 208	80	29.1	Dubna	80	51.3
	25	14.9		25	35.1
	8	4.7		8	26.6
	0.8	2.5		0.8	5.5
EJ 260	80	44.8	Bicron	80	45.5
	25	15.5		25	39.5
	8	14.3		8	11.5
	0.8	6.6		0.8	8.7

Table 1. Table showing the % transmission loss for all samples at a wavelength of 420 nm, over the different doses

Dose	Sample	%Trans. diff. (Day1 - 1 Week)
25 MGy	EJ 200	30.38
	EJ 208	37.06
	EJ 260	6.45
		%Trans. diff. (Day 1 - 4 Weeks)
8 MGy	EJ 200	15.05
	EJ 208	5.92
	EJ 260	2.22

Table 2. Table showing the % transmission difference for different days over two doses

In Figure 3 as each sample is exposed to radiation, the absorptive tint evident in the unirradiated graph tends to shift towards longer wavelengths as the dose is increased. In addition, a decline in percentage transmission is evident as well and the percentage transmission loss increases with increase in dose. Also, Figure 4 shows that indeed the samples do undergo some form of recovery after the radiation process. Between the day of irradiation and 2 or 3 days after irradiation,there is a a substantial amount of increase in percentage transmission, indicating the sample recovery. The Following weeks also indicate some healing, but small in comparison to the first few days following the irradiation process. The recovery was also evident in the visual changes of the sample irradiation spots. For very high doses, the dark yellow radiation spot faded to a normal yellow and for the lower doses the spot was nearly invisible even on the days of irradiation.

Table 1 shows the percentage loss in transmission for each sample over each dose in comparison to its unirradiated. That is, for each dose at a wavelength of 420 nm, the percentage transmission loss was calculated as follows:

$$\%\ transmission\ loss = \%\ transmission\ of\ unirradiated - \%\ transmission\ of\ irradiated \quad (3)$$

From the table, one can see that the PS based plastic scintillators (protvino and dubna) undergo the most decrease in transmission for most of the doses. Also, the bicron shows high percentage loss in transmission for all the doses. On the other hand, the PVT based scintillators provided by ELJEN technologies show the least amount of decline in percentage transmission for all the doses in comparison to all the other grades. On the other hand, Table 2 shows the difference in % transmission between two different days that transmission data was taken. This was done for all the EJ 200, EJ 208 and EJ 260 for only two specific doses. For the higher dose EJ 208 showed the most amount of healing as a large difference in transmission is seen. For the smaller dose, EJ 200 reflected the most amount of healing in comparison to the others.

6. Conclusions

Even though plastic scintillators generally possess overall commendable properties, their vulnerability when exposed to high doses of radiation is a major predicament. Unfortunately, all types of scintillators are affected by radiation. Studies done by Bross [2] indicate that light yield losses in plastic scintillators are not due to the dopant degradation, but rather to the damage within the base matrix. In addition, according to Torrisi [7], when PVT based polymers are exposed to high radiation doses, their chemical bonds are broken and their polymer properties are changed. Apparently, at the breaking points of the carbon and hydrogen bonds, free radicals are formed and hydrogen degassing occurs, resulting in a material which is rich in carbon. The formation

Journal of Physics: Conference Series **645** (2015) 012021 doi:10.1088/1742-6596/645/1/012021

of free radicals is bound to affect the scintillating mechanism of the plastic scintillators as they tend to absorb fluorescent light thus limiting the light yield of the scintillators.

In our transmission results, we found that indeed an increase in radiation results in a decline in light transmission for all of the different grades. This could be because the dopants are no longer emitting the base scintillation but are rather absorbing it instead. When comparing the transmission spectra of each dose with its unirradiated in the wavelength range that is read out by the optical fibres of the ATLAS detector, we found that the PVT based scintillators provided by ELJEN technologies showed the least amount of decline in percentage transmission in comparison to the other scintillators. This is evident in the data in Table 1. Amongst the PVT based scintillators, we found that EJ 208 showed the least amount of decline in percentage transmission for most of the doses. On the other hand, EJ 260 showed the highest amount of decline in percentage transmission in comparison to the other two brands.

Also, the annealing of a sample after the process of irradiation plays an important role in establishing whether a material is radiation hard or not. From the transmission data collected on different days in order to observe annealing, we found that the plastic scintillators do indeed undergo some healing after radiation. From comparing the healing of two different doses we found that EJ 208 and EJ 200 both experienced a substancial amount of healing in comparison to EJ 260. Also, it was also noted that for all the grades, the most amount of healing occurred within the first few days after irradiation in comparison to weeks later.

Moreover, from these results and observations, it is clear that even though EJ 200 showed the best transmission properties at the start, after irradiation, EJ 208 reflected more radiation hardness. This is supported by the fact that it showed the least amount of decrease in percentage light transmission as well as also healed a substancial amount in a short period of time. Light yield results are currently being done and will further aid in the recommendation of the best plastic scintillator grade.

References

[1] A. Kastalsky, S. Luryi, B. Spivak (Aug 25 2006)., "Semiconductor High-Energy Radiation Scintillator detector", Nucl.Instrum.Meth.A565:650-656,2006. [arXiv:physics/0608248 [physics.ins-det]]

[2] A.D. Bross (Nove 14 1988), "Radiation damage in scintillators", *Proceedings of the Workshop on Scintillating fiber Detector Development for the SSC*, Fermi National Accelerator Laboratory, Batavia, IL, 741-768

[3] ATLAS EXPERIMENT, retrieved on 10 September 2014 from `en.wikipedia.org/wiki/ATLAS-experiment`

[4] A. Succurro (Sep 14 2011), The ATLAS Tile Hadronic Calorimeter performance in the LHC collision era, Phys.Procedia 37 (2012) 229-237 ATL-TILECAL-PROC-2011-009,ATL-COM-TILECAL-2011-025

[5] ELJEN technology, "Organic scintillators for timorous tech". Retrieved on 10 October 2014 from `www.eljentechnology.com`

[6] K. D. Rakes (2008), Evaluating the response of polyvinyl toluene scintillators used in portal detectors, Airforce Institute of Technology Wright-Patterson. Retrieved on 14 September 2014 from `www.dtic.mil/dtic/tr/fulltext/u2/a483290.pdf`

[7] L. D. Torrisi (2002), "Radiation damage in Polyvinyl toluene (PVT)", *Radiation Physics and Chemistry*, 63, 89-92

Journal of Physics: Conference Series **645** (2015) 012022 doi:10.1088/1742-6596/645/1/012022

PGAS in-memory data processing for the Processing Unit of the Upgraded Electronics of the Tile Calorimeter of the ATLAS Detector

Daniel Ohene-Kwofie and Ekow Otoo

School of Electrical & Information Engineering, University of the Witwatersrand, Johannesburg, South Africa

E-mail: daniel.ohene-kwofie@cern.ch

Abstract.
The ATLAS detector, operated at the Large Hadron Collider (LHC) records proton-proton collisions at CERN every 50ns resulting in a sustained data flow up to PB/s. The upgraded Tile Calorimeter of the ATLAS experiment will sustain about 5PB/s of digital throughput. These massive data rates require extremely fast data capture and processing. Although there has been a steady increase in the processing speed of CPU/GPGPU assembled for high performance computing, the rate of data input and output, even under parallel I/O, has not kept up with the general increase in computing speeds. The problem then is whether one can implement an I/O subsystem infrastructure capable of meeting the computational speeds of the advanced computing systems at the petascale and exascale level.

We propose a system architecture that leverages the Partitioned Global Address Space (PGAS) model of computing to maintain an in-memory data-store for the Processing Unit (PU) of the upgraded electronics of the Tile Calorimeter which is proposed to be used as a high throughput general purpose co-processor to the sROD of the upgraded Tile Calorimeter. The physical memory of the PUs are aggregated into a large global logical address space using RDMA- capable interconnects such as PCI-Express to enhance data processing throughput.

1. Introduction

The Large Hadron Collider, is the most powerful proton-proton collider ever built [1, 2]. The discovery of the Higgs Boson [3, 4] was independently observed by the ATLAS and Compact Muon Solenoid (CMS) detectors. A Toroidal Large Hadron Collider Apparatus (ATLAS) is the largest of all the LHC detectors. It consists of a series of concentric rings: the inner detector, Tile Colorimeters and the Muon Spectrometers. TileCal is the central hadronic calorimeter of the ATLAS experiment at the LHC at CERN. It is primarily used to measure the energy and direction of hadrons and jets as they are produced. The Trigger and Data Acquisition System (TDAQ) is designed for event selection, processing and storage of the read-out data of the detector [5]. This selection mechanism is based on three trigger levels in the data flow that defines the different domains for the read-out electronics in terms of methods and rates for this selection.

Detector electronics are being upgraded (as part of scheduled upgrade phases on the accelerator and experiments components) to allow an increase in the Level-1 acceptance rate of events from 70kHz to 100kHz [6]. There by subsequently increasing the raw data sent for processing. These upgrades are aimed to extend the physics reach of the LHC measurement programme.

1.1. The Off-line Data processing Challenge

Scheduled upgrades to the ATLAS detector anticipated in 2022 will result in a much higher rate of collisions [7] at the LHC, resulting in an increase by 200 times the current rate to over 41Tb/s data output from the TileCal [5]. Storing such massive dataset for off-line processing presents a

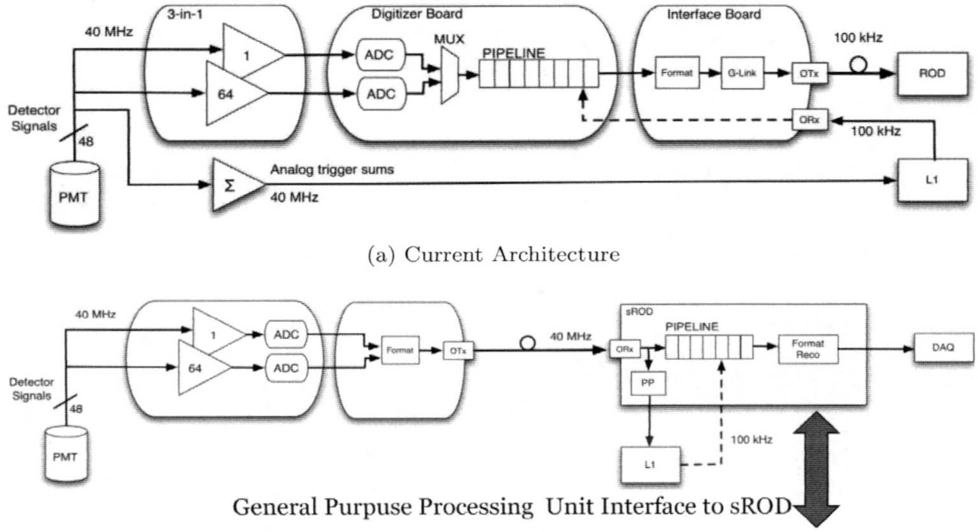

(a) Current Architecture

(b) Upgraded Architecture

Figure 1: ATLAS Tile Calorimeter

great challenge and is not desirable. The MAC project at the University Of Witwatersrand is aiming for a cost-effective, and high data throughput Processing Unit (PU), using several consumer ARM processors in a cluster configuration, as general purpose co-processor to augment the read-out system (sROD) of the upgraded TileCal.

A major challenge with processing such large volumes of data is the inpt/output (I/O) sub-system. The continuously growing gap between CPU and I/O speed has resulted in the conspicuous performance gap between the processor speeds and what the storage I/O subsystem can deliver. The solution is to reduce disk accesses and enhance in-memory data processing.

We present a brief overview of a complete in-memory storage system for on-line data processing in the ARM cluster configuration for the upgraded electronics of the TileCal. In-memory data processing provides extremely fast response time and very high throughput, with an average of about $100 - 1000$ times lower latency for a complete Random Access Memory(RAM) storage than disk-based storage systems and consequently a $100 - 1000$ times greater throughput [8].

2. PGAS Architecture

The architecture consists of a cluster of ARM processing units, part of whose memories are aggregated to form a combined logical global address space. The general operational architecture includes low cost ARM processing units interconnected via PCI Express interconnects (PCIe). The PCIe interconnect facilitates Remote Data Memory Access (RDMA) and offers very low-latency host-to-host transfers by copying the information directly between the host application memories. This enhances a seamless global logical address space for in-memory data processing as a low-overhead protocol. They are relatively affordable, low power and also provide straightforward, standards-compliant extensions that address multi-host communication and I/O sharing capabilities. Figure 2a illustrates the schematic diagram of the architecture. Part of the on-going research work is to address a number of challenges related to fault-tolerance and high speed access of the memory resident data.

3. The PGAS High Data Throughput Architecture

The Figure 2 and the experiments conducted so far only demonstrates the processing capability of the PU under the PGAS computing model. This is easily transformed into a configuration where the aggregated memories of the ARM processors serve as a large in-memory buffer for data staging. By augmenting the ARM cluster with data input ports and data output ports to external hierarchical storage devices, data can be streamed into the ARM cluster memory using parallel I/O. This in-memory data can then be compressed and streamed out using parallel I/O to a first level of attached solid state disks (SSDs) and subsequently to archival disks. We envisage that the high processing speed of compressing the data and streaming it out to other high speed devices or memory of other devices, will outweigh the rate that data is streamed into the ARM cluster. Experiments are underway to test this architecture.

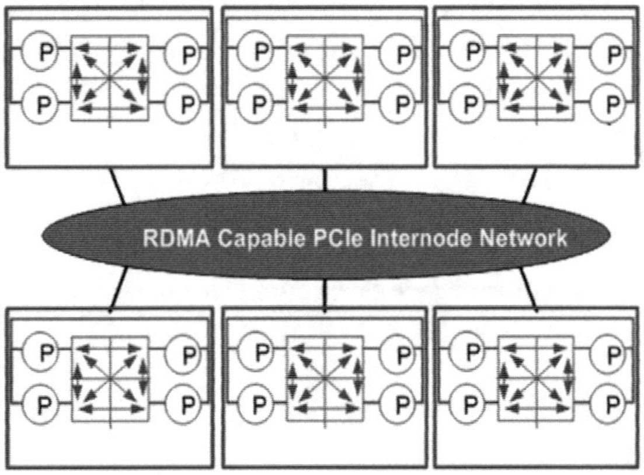

(a) Schematic Diagram of the Architecture

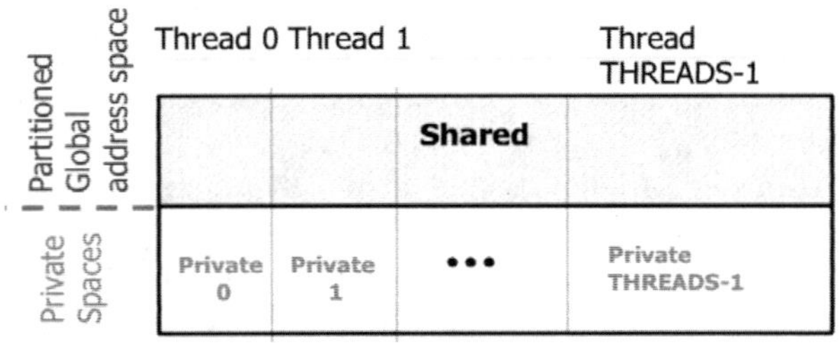

(b) PGAS Memory layout

Figure 2: Schematic diagram showing operational layout of the architecture (a). (b) shows the memory layout indicating the global shared memory as well as the private address space. Data is distributed to enhance locality and thus ensure greater throughput.

4. Preliminary Evaluations

Preliminary investigations conducted show promising results. We benchmark the system with the NASA Advanced Supercomputing (NAS) Parallel Benchmarks [9]. This benchmark was designed not just for parallel-aware algorithmic and software methods but also to provide an easy verifiability of correctness of results and performance figures. The evaluations are run using the Fast Fourier Transform(FFT) algorithm which solves a 3D partial differential equation using an FFT-based spectral method [9], also requiring long range communication. FFT performs three one-dimensional(1-D) FFT's, one for each dimension.

The evaluation was done on 4 nodes of the Wits Tegra k1 (2.3GHz Quad-Core ARM Cortex-A15) cluster with 2GB of memory each and 1Gbp Ethernet interconnect between nodes.

We ran the benchmark with varying workloads as well as varying number of threads. Each workload is ran 6 times and the resulting throughput in MFLOPS (floating point operations per second) reported. A maximum of 4 threads per node are spawned and the dimensions of the FFT is varied from small ($64 \times 64 \times 64$ 3D grid) to large ($256 \times 256 \times 128$ 3D grid). The User Datagram Protocol(UDP) is used as the inter-process communication protocol between cluster nodes. UPC supports UDP, Message Passing Interface (MPI) as well as RDMA capable interconnects(e.g infiniband, PCIe, 10GbE, etc.) for inter-node communication. The current experimental setup did not have RDMA and therefore UDP was used since it is usually faster than MPI as far as inter-node communication over Ethernet is concerned.

Figures 3a and 3b show the results for the data processing throughput in MFLOPS with varying number of threads and workloads. Generally, as the number of threads increase, there is a corresponding increase in the data processing throughput since less processing is done per thread. We observe a significant and better performance increase with PGAS as depicted in the figures (about 3× more throughput than the NON-PGAS FFT).

Figure 4 shows the average latencies for each run of the experiment. There is a general increase in latency as workload size increases as expected. We also observe that as the number of threads increase latency drops a little and begins to increase after 8 threads. This is due to increase in inter-node communication. PGAS performs much better with lower latencies as compared to its NON-PGAS variant.

5. Conclusions

Management, efficient access and analysis of the Petabytes of data, that is likely to be generated and/or used in the upgraded ATLAS TileCal present extremely challenging tasks. I/O bottlenecks in processing such huge amounts of data require techniques that utilise higher levels of the memory hierarchy to enhance data throughput.

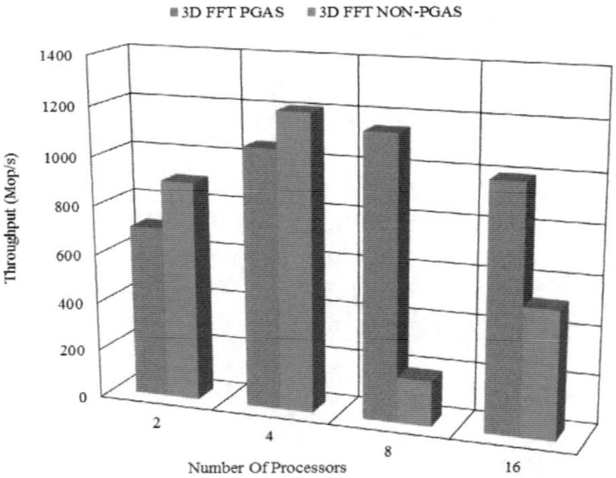

(a) small workloads

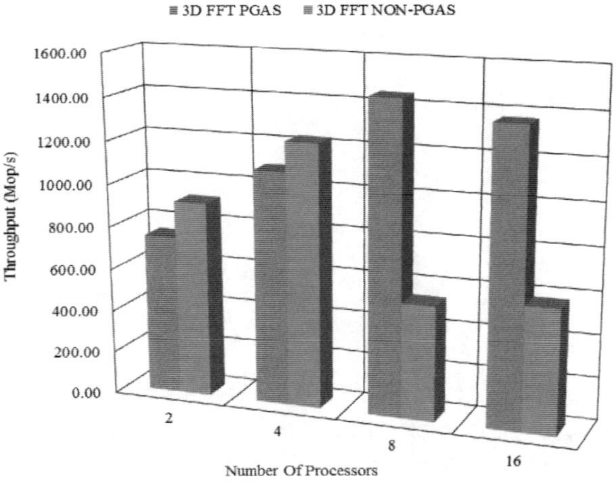

(b) large workloads

Figure 3: Data Throughput

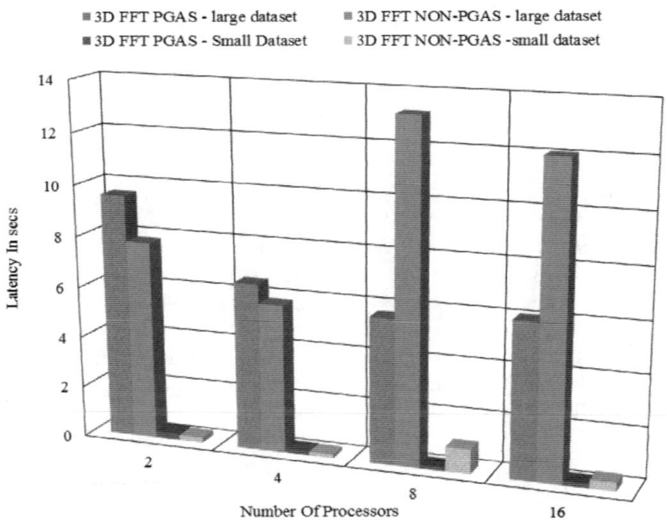

Figure 4: Average latencies for varying workloads

The PGAS model presents an efficient way to process data in a distributed environment by providing a global partitioned shared address space for in-memory data. This enhances the data processing throughput . Additionally, the use of low cost CPUs such as ARM or Atom PUs with PCIe interconnects, ensure low power consumption (high performance per watt) and thus cost effective alternative for data processing in ATLAS TileCal.

Future work anticipated includes further rigorous experimentation using the ARM PUs with RDMA capable PCIe intra/interconnects for kernel bypass applications.

References

[1] CERN Communication Group, "The CERN LHC faq Guide." `http://cds.cern.ch/record/1165534/files/CERN-Brochure-2009-003-Eng.pdf`, February 2009.

[2] CERN, "The Large Hadron Collider." `http://home.web.cern.ch/topics/large-hadron-collider`, 2014.

[3] J. T. Baines and et al., "An overview of the atlas high-level trigger dataflow and supervision," *Nuclear Science, IEEE Transactions on*, vol. 51, pp. 361–366, June 2004.

[4] R. Reed, F. Carrio, P. Moreno, C. Solans, J. Souza., and A. Valero, "A revised high voltage board for the consolidation of front end electronics on the tile calorimeter of the atlas detector at the lhc," in *South African Institute of Physics Conference,SAIP 2013 Poster Presentation*, (University Of the Witwatersrand, Johannesburg, South Africa), Jul 2013.

[5] F. Carrioa and et. al, "The sROD module for the ATLAS Tile Calorimeter Phase-II Upgrade Demonstrator," *Journal of Instrumentation, JINST*, vol. C02019, no. Issue 09, 2014.

[6] The ATLAS Collaboration, "Letter of Intent for the Phase I Upgrade of the ATLAS." `https://cdsweb.cern.ch/record/1402470`, 2012.

[7] The ATLAS Collaboration, "Letter of Intent for the Phase II Upgrade of the ATLAS." `https://cdsweb.cern.ch/record/1502664`, 2012.

[8] J. Ousterhout, P. Agrawal, D. Erickson, C. Kozyrakis, J. Leverich, D. Mazières, S. Mitra, A. Narayanan, D. Ongaro, G. Parulkar, M. Rosenblum, S. M. Rumble, E. Stratmann, and R. Stutsman, "The Case for RAMCloud," *Commun. ACM*, vol. 54, pp. 121–130, July 2011.

[9] O. Serres, N. Andreev, C. Francois, A. Abhishek, A. Smita, B. Veysel, Y. Yiyi, S. Chauvin, F. Vroman, and T. El-Ghazawi, "UPC NAS Parallel Benchmarks." `http://threads.hpcl.gwu.edu/sites/npb-upc`, April 2011.

Journal of Physics: Conference Series **645** (2015) 012023 doi:10.1088/1742-6596/645/1/012023

Characterization of plastic scintillators using magnetic resonance techniques for the upgrade of the Tile Calorimeter in the ATLAS detector

C. Pelwan, H. Jivan, D. Joubert, J. Keartland, S. Liao, G. Peters, E. Sideras-Haddad

School of Physics, University of the Witwatersrand, Johannesburg 2050, South Africa

E-mail: chad.pelwan@gmail.com

Abstract. In this study we look at radiation damage and its adverse effects on plastic scintillators housed within the Tile Calorimeter (TileCal) of the ATLAS detector. The study focuses on determining how the interaction of ionizing radiation with plastic scintillators effects their efficacy and desired properties such as high light output and fast decay time. Plastic scintillators form an integral part of the ATLAS trigger system and their optimal functionality is paramount to the success of ATLAS. Electron paramagnetic resonance (EPR) provides insight into the electronic structure of the plastics and can characterize the damage caused by ionizing radiation. Density functional theory (DFT) calculations will be performed in order to simulate the EPR signal. Preliminary EPR results investigate four different types of plastic scintillators. These include three polyvinyl-toluene based Eljen technologies: EJ200, EJ208 and EJ260, and one polystyrene based Dubna sample. It has been observed that the Dubna sample, identical on the current scintillator used in the ATLAS detector , undergoes more structural damage when compared to the Eljen samples.

Introduction

The increase in luminosity to $\sqrt{s} = 13$ TeV at the start of Run 2 in April 2015 will require that the Large Hadron Collider (LHC), its four detectors, and various components be upgraded in the phase one upgrade in 2018 [1].

The minimum bias trigger scintillator (MBTS) plastics are situated around the beam pipe and have a pseudo-rapidity, η of $2.07 \leq |\eta| \leq 3.84$ [2]. There are 16, 2 centimetre thick, polystyrene based plastics on each side of the ATLAS detector. The plastics are manufactured in Dubna and doped with 2% organic dopants optimize their functionality. Their main purpose, before nominal operation of the LHC, was to veto background from the beam and, since an effective mean was needed for proton-proton collisions, they provided a vital role in providing this. When the MBTS plastics interact with energetic particles, they produce light that is passed down to photo-multiplier tubes (PMTs) via wavelength shifting fibres where they are converted into a digital signal and analysed. However, these plastics are susceptible to radiation damage induced by the energetic diffractive particles that interact with them and it has been found that this interaction causes bonds to break in the benzene ring which is found in the molecule and is responsible for the scintillation mechanism in the plastics [3].

Journal of Physics: Conference Series **645** (2015) 012023 doi:10.1088/1742-6596/645/1/012023

This study investigates the effect ionizing radiation has when it interactions with the various types of plastic scintillators. Electron paramagnetic resonance (EPR) investigates unpaired electrons and ions that are formed due to structural damage [4]. In addition to this, an ab initio computational approach will be used in order to replicate the EPR spectra seen by isolating the benzene derivative molecules responsible for scintillation the plastic scintillators.

Electron Paramagnetic Resonance

EPR spectroscopy is a powerful technique for studying unpaired electrons and ions in samples. Experimental results have shown that, as the energetic particles ionize the plastics, C-H bonds in the benzene derivative molecules break [5]. Therefore, employing EPR will allow us to gain vital information about the electronic structure of the samples. When as sample is placed between a large, homogeneous magnetic field, the magnetic moments of the unpaired electrons and ions couple to the magnetic fields. A resonance frequency, ω, is then applied and corresponds to a transition between the two spin states. An absorption energy can be seen described by the resonant condition:

$$\Delta E = h\omega = g\mu_B B_0, \tag{1}$$

where μ_B is the Bohr magnetron, B_0 is the external magnetic field intensity, and g is called the g-factor.

Materials

Replicating the harsh environment the scintillators experience within the ATLAS detector is a non-trivial task. Two types of scintillators were considered for this study: one polyvinyl-toluene based sample with three candidates of Eljen technology and one polyester based sample. The three Eljen samples are named EJ200, EJ208 and EJ260. The Dubna sample is identical to the MBTS plastic used in the ATLAS detector. Each of these samples contain their respective dopants that aid the scintillation mechanism.

The tandem accelerator at iThemba Laboratory for Accelerator-Based Sciences (LABS) provided energetic proton beams that irradiated the samples. The 3 MV tandem accelerator is capable of energizing protons to around 6 MeV. This is sufficient to replicate the radiation environment seen by the MBTS plastics in the ATLAS detector. The particle interaction with the material was modelled using SRIM (Stopping and Range of Ions in Matter) which predicted that the stopping range in the two types of scintillators which occurred around 470 μm for the 6 MeV protons[6]. TRIM (Transport of Ions in Matter), an extension of SRIM, was used to calculated the average energy of each impinging proton at around 2.07 MeV. This value was needed to calculate the total absorbed dose that the scintillator experiences by the protons. Before this could be done, each sample needed to be machined to a size that would fit onto a carousel placed into the tandem accelerator and a width less than the stopping range. A volume of 1 cm $\times$ 1 cm $\times$ 250 μm was chosen and each sample was cut and polished to match this volume. Each plastic type was irradiated to two doses: 0.164 MGy and 1.46 MGy where Gy is the absorption of a joule of radiation energy per kilogram [7]. The absorbed dose was calculated using

$$D = \frac{I_b \; A \; E_{ave} \; \Delta t}{q \; m_{ir}} \; \text{Gy}, \tag{2}$$

where I_b is the beam current, A is the area of the irradiated spot, E_{ave} is the average energy per proton, Δt is the time spent irradiating the sample, q is the charge of the proton, and m_{ir} is the mass of the irradiated spot. After each sample has been irradiated they (and the un-irradiated samples) can undergo EPR testing. EPR experiments were performed at the University of Witwatersrand and have verified this assumption as a signal from both un-irradiated and irradiated samples were seen.

Journal of Physics: Conference Series **645** (2015) 012023 doi:10.1088/1742-6596/645/1/012023

EPR Results

The following results show the peak intensities of different samples. The peak heights are in arbitrary units as because they depend on sample size. However, since each sample has the same density and volume we do not need to apply a mass normalization. The resonant frequency is kept constant and the magnetic field is swept. Absorption peaks form at different magnetic field intensities depending on the which type ion is being detected.

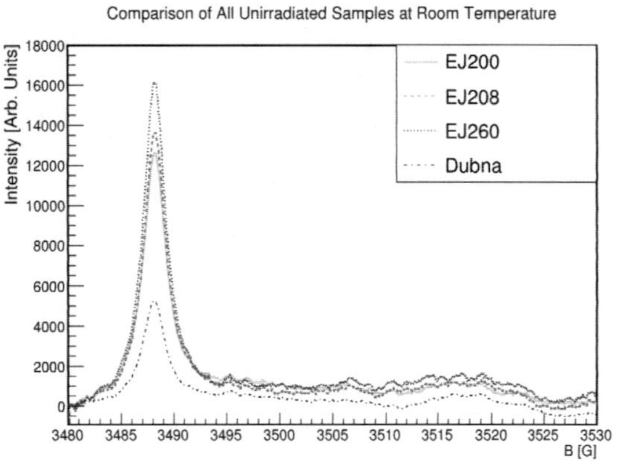

1 (a)

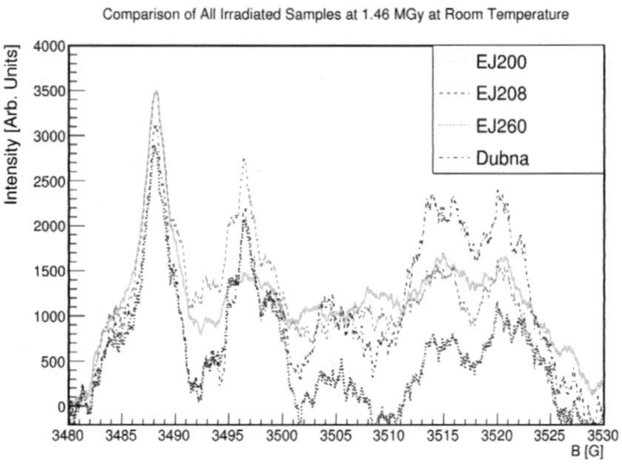

1 (b)

Figure 1: Comparison of all sample intensities as magnetic field is swept for un-irradiated samples (a) and irradiated at a dose of 1.46 MGy (b).

Un-irradiated samples seen on the left of figure 1 (a) show that unpaired electrons and ions exist in the sample before irradiation. This is indicated by the absorption observed around 3488 Gauss which corresponds to the intensity of unpaired electrons in a sample.

Figure 1 (b) clearly indicates secondary peaks begin to form as samples are irradiated. The interaction with the ionizing radiation causes a formation of new unpaired electrons and ions ultimately damaging the scintillation mechanism as it alters the structure of the scintillator.

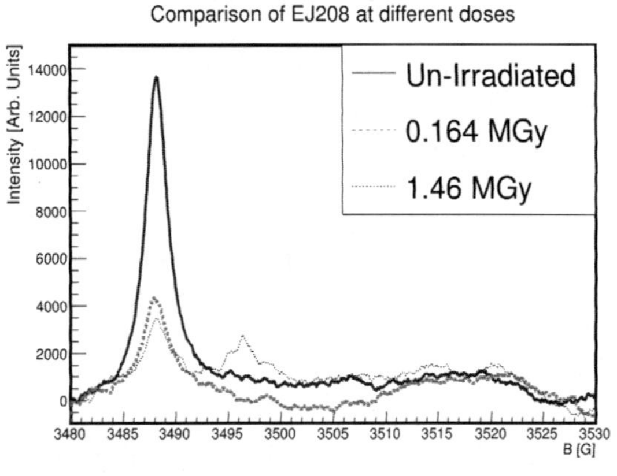

2 (a)

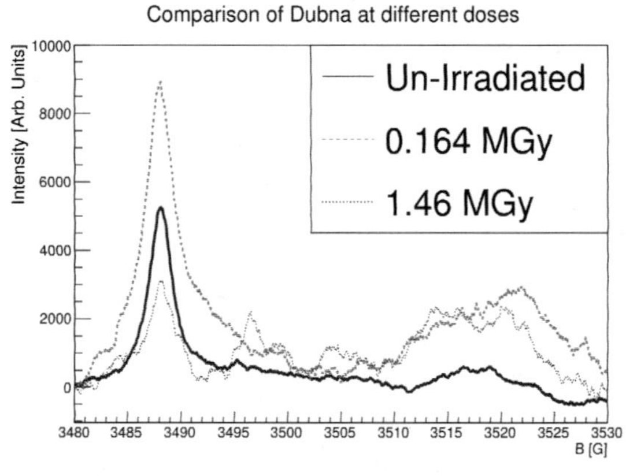

2 (b)

Figure 2: Comparison of polyvinyl-toluene sample EJ208 (a) and polystyrene sample Dubna (b) at various doses as magnetic field is swept for a constant resonant frequency.

Figure 2 (a) indicates the change is spectra of a polyvinyl-toluene sample, EJ208 and polystyrene Dubna sample (figure 2(b)) as dose is increased inducing more damage. It is evident that, initially, there is a larger intensity of unpaired electrons in EJ208. As the samples are bombarded by protons this intensity decreases. This is could be due to the C-H bonds breaking, releasing electrons and unpaired hydrogen atoms in the sample.

Future Work

Computation Replication of Results

EPR spectroscopy is a characterization technique and give a qualitative view of the scintillators and how damage alters their structure. A quantitative approach is needed in order to fully understand the problem. Thus, we look to model the system using a computational ab-initio approach. This will be done using density function theory (DFT) which is a theory for a correlated many-body systems that views any property of the system as a functional ground state energy, $E_0(\vec{r})$ [8]. From the ground state functional properties about the electronic structure of the system can be derived.

By isolating the optically active molecule (the benzene ring and its functional groups responsible for the scintillation mechanism) for both polyvinyl toluene and polystyrene, we can use DFT methods to simulate the EPR [9]. The goal would then be to correlate the two spectra sets with properties from the DFT predictions. These include, but are not limited to, number of unpaired electrons and ions in the system, ground and excited states of the system, and identification of functional groups and ions in un-irradiated and irradiated samples.

Conclusion

This project, accompanied with various other studies done on the plastic scintillators aims to find a suitable candidate for the replacement of the MBTS plastics as well as the plastic scintillators housed inside the gap region of the TileCal once they have lost their efficacy. We would also have a deeper understanding of radiation damage in plastic scintillators as a whole and hopefully this knowledge would aid in the design and manufacture of a plastic scintillator that is less susceptible to radiation damage than the technology currently available.

References

[1] G Aad and Collaboration A 2008 *INSTITUTE OF PHYSICS PUBLISHING AND SISSA* **08003**
[2] Kwee R, Kolanoski H and Ellis N 2012 *Development and Deployment of an Inner Detector Minimum Bias Trigger and Analysis of Minimum Bias Data of the ATLAS Experiment at the Large Hadron Collider* Ph.D. thesis Humboldt-Universitat zu Berlin
[3] Huston J 2012 *Michigan State University* 12
[4] Weil J and Bolton J 2007 *Electron Paramagnetic Resonance* (John Wiley & Sons, Inc., Hoboken, New Jersey) ISBN 9780471754961 URL `http://www.wiley.com/go/permission.`
[5] Torrisi L 2002 *Radiation Physics and Chemistry* **63** 89–92 ISSN 0969806X
[6] Ziegler J F 2013 SRIM - The Stopping and Range of Ions in Matter URL `http://srim.org`
[7] BIPM 2006 *Organisation Intergouvernementale de la Convention du Mètre* **1** 91–180
[8] Martin R M 2006 *Understanding Carbon Nanotubes* **1914** 199–276 ISSN 0163-1829 URL `http://dx.doi.org/10.1007/3-540-37586-4_4`
[9] Marx D and Hutter J 2009 *Ab Initio Molecular Dynamics* (CAMBRIDGE UNIVERSITY PRESS) ISBN 9780521898638 URL `www.cambridge.org/9780521898638`

Journal of Physics: Conference Series **645** (2015) 012024 doi:10.1088/1742-6596/645/1/012024

An Integration Framework Tool for ATCA Chassis in the ATLAS Detector Control System

Robert Graham Reed

School of Physics, University of the Witwatersrand, Johannesburg 2050, South Africa

E-mail: `robert.reed@cern.ch`

Abstract. The Large Hadron Collider at CERN is scheduled to undergo a major upgrade in 2022. The ATLAS collaboration will do major modifications to the detector to account for the increased luminosity. More specifically, a large proportion of the current front-end electronics, on the Tile Calorimeter sub-detector, will be upgraded and relocated to the back-end. A Demonstrator program has been established as a proof of principle. A new system will be required to house, manage and connect this new hardware. The proposed solution will be an Advanced Telecommunication Computing Architecture (ATCA) which will not only house but also allow advanced management features and control at a hardware level by integrating the ATCA chassis into the Detector Control System.

1. Introduction

The Large Hadron Collider (LHC) collides protons at extremely high energies in order to help answer questions about the sub atomic universe. The ATLAS detector [1] is one of two general purpose detectors which records over 600 million collisions a second. The ATLAS detector is scheduled to undergo a major upgrade in the year 2022 [2]. The ATLAS detector is sub-divided into numerous sub-detectors, each one having a specific function. The Tile Calorimeter (TileCAL), the central region, is used to measure energies and directions of hadrons, jets, τ and leptons. Plastic scintillators which, when particles pass through, emit light to photo-multiplier tubes (PMTs) are used to measure energy and direction of hadrons. These light signals are directed to front-end electronics where they are digitised and processed for the first level of triggering.

Data from the front-end electronics is transferred off the detector to the back-end electronics where further processing occurs. These electronics are house in Versa Module Europa (VME) crates. These systems are over 30 years old and will not be able to sustain the increased data rates that will be required in the upgrade of the LHC in the year 2022. As a result the entire front and back-end electronics will be upgraded for the Phase II upgrade [2].

2. Super Read Out Driver

A part of the Phase II upgrade is the redesign of the Read Out Driver (ROD) of the TileCalorimeter. This will involve electronics from both the front and back-end system. This new improved board, called the Super ROD (sROD), will contain the sensitive electronics that are currently housed in the front-end area. This will allow the full data rate of 40 MHz to exit the detector and be processed in the back-end area. Figure 1 shows the schematics of the upgrade with the pipelines and triggering contained in the new super ROD (sROD). Since the sROD is housed in the back-end it will be accessible during run time allowing easy maintenance and troubleshooting. Table 1 shows the expected bandwidth that will be provided by the new sROD and the TileCAL as a whole [3].

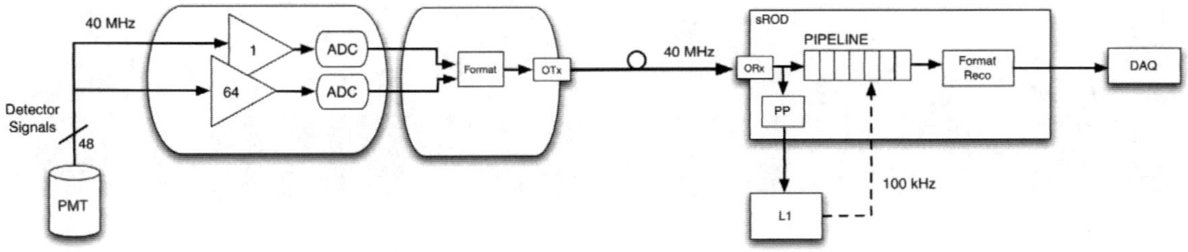

Figure 1: Upgraded read out electronics.

The sROD will be housed in an Advanced Telecommunications Computing Architecture (ATCA) chassis. The ATCA standard will be replacing the previous VME standard as it is superior in many aspects [4]. Once the sROD prototypes have been tested the designs will be brought to South Africa where the PCB will be manufactured and all components mounted. The sROD will be inserted into the ATCA system at the University of the Witwatersrand and tested.

3. ATCA Framework

The ATCA chassis allows intelligent monitoring and control while offering high speed connectivity via a 40 Gbps backplane. Figure 2 shows a front view of the chassis that is currently installed at the University of the Witwatersrand. This model has six slots in a dual star topology offering redundancy. A 10 Gbps switch module provides the routing and switching configuration. A carrier board allows smaller cards, called Advanced Mezzanine Cards (AMC), to be inserted. The sROD is a doubled sized AMC card that can be inserted as shown in Fig 2 (b). All boards can be inserted and extracted while the system is powered on. This is called Hot Swapping. The Shelf Manager is responsible for the all the axillary services such as power control, fan speeds, temperature readings and voltage readings among others. Communication to the Chassis is done via a Simple Network Management Protocol (SNMP). This protocol is used since the Detector Control Software, WinCC, has a built in driver to manage the connection. The ATCA follows the PCI Industrial Computer Manufacturers Group (PICMG) standard. PICMG, as defined on their website, is "a consortium of companies who collaboratively develop open specifications for high performance telecommunications and industrial computing applications." [5].

Table 1: Tile Calorimeter Bandwidth Upgrade.

Phase	Present	Upgrade
Number of fibers	256	4096
Fiber Bandwidth	800 Mbps	10 Gbps
Total Bandwidth	205 Gbps	41 Tbps

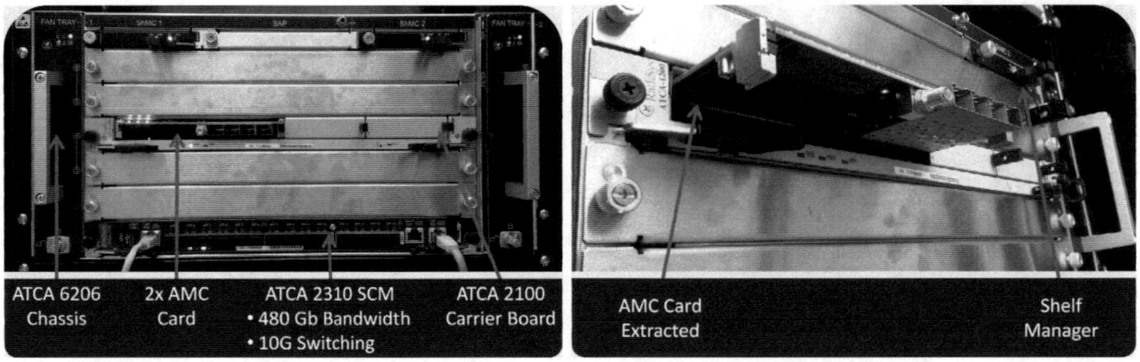

(a) (b)

Figure 2: a) Front of ATCA chassis and b) AMC extracted from chassis at Wits.

4. The ATLAS Detector Control System

The Phase II upgrade will issue in new electronics such as the ATCA back-end infrastructure. This new hardware needs to be integrated into the existing Detector Control System (DCS) [6]. The DCS is a highly parallel system with distributed nodes that monitor and control 12 individual sub detectors in a redundant fashion. The DCS has the task of bringing the detector and all components into various desired operational states while monitoring and performing corrective actions in case of abnormal behaviour, all in an automated fashion.

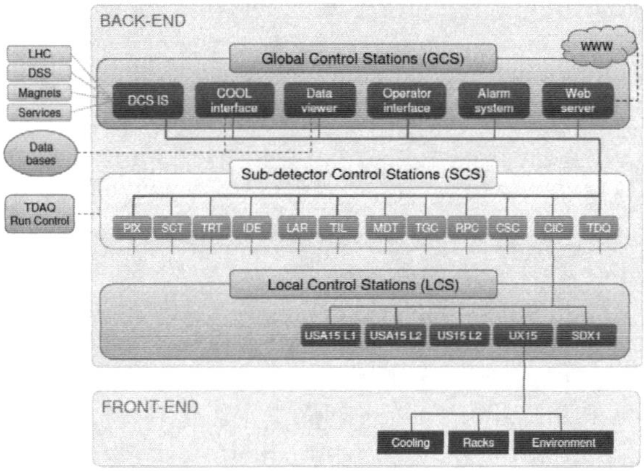

Figure 3: Schematic of the ATLAS Detector Control System [6].

Figure 3 shows the distributed nature of the DCS and the logical separation of the controlling nodes. It consists of various drivers arranged in tiers according to according to geographic, function or dependency groups. The software is called WinCC Open Architecture which is a product of Siemens [7].

All sensors and variables of interest in the ATCA system need to be polled in a regular fashion to allow the DCS to correctly deduce the state of all devices. This requires the creation of data points in the WinCC software that is very specific to the hardware in question. A framework tool has been designed which provides the expert the means to automatically generate the required data points. It does this with all the correct configurations such as the SNMP address, formatting, descriptions, polling intervals, access permission and more. This framework tool is

Journal of Physics: Conference Series **645** (2015) 012024 doi:10.1088/1742-6596/645/1/012024

called fwATCA.

5. fwATCA Framework Tool

The fwATCA framework tool is required to perform four functionalities:

(i) Search - This is the most important. The framework needs to search through the ATCA Shelf and determine what it contains and find all the sensors that correspond to each device.

(ii) Sort - Once all the devices have been found, along with their sensors, the information needs to be sorted in a meaningful and clear way.

(iii) Create - The sorted information must then be generated into data points that are accessible by the Detector Control System.

(iv) Configure - After creation one may want to configure a selected data point (or all) with descriptions, alerts, formatting, polling intervals or simply to delete them.

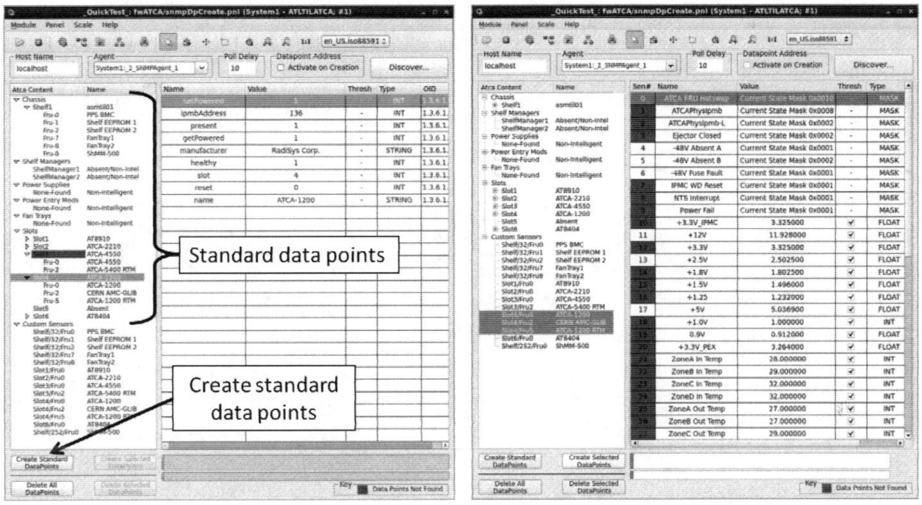

Figure 4: Framework panel that allows data point creation.

Figure 4 shows the framework panel that has been able to search through the ATCA shelf. Sort the information into two main categories, with sub categories, and then create the data points (The shaded colour around the sensor name indicates that the sensor has not been created). The sorting of the data points is defined by two categories: Standard Data Points and Custom Data Points. The first being sensors that are populated by the Shelf Manager. Standard sensors are always available as they belong to the chassis itself and are always present. The Standard Data Point category is then broken into smaller groups depending on the physical devices itself such as fan trays, power modules, shelf managers or physical slot bays. The number of devices in each category will depend on the chassis type since some chassis only have six slots while others may have fourteen. The second category, Custom Data Points, contains the sensors that belong to the boards that are inserted. Naturally these sensors are only available if the board is present. The custom sensors are provided by the board manufacturer and must meet the PICMG specifications. The information these sensors provide is different for each board which means they are harder to integrate in an automated fashion. However, this has been achieved. The custom sensors are processed after creation allowing them to be categorised according to the information they provide.

Figure 5 provides the expert with the tools requires to configure the data points after creation. This involves creating descriptions, viewing (or deleting) data points that are no longer needed

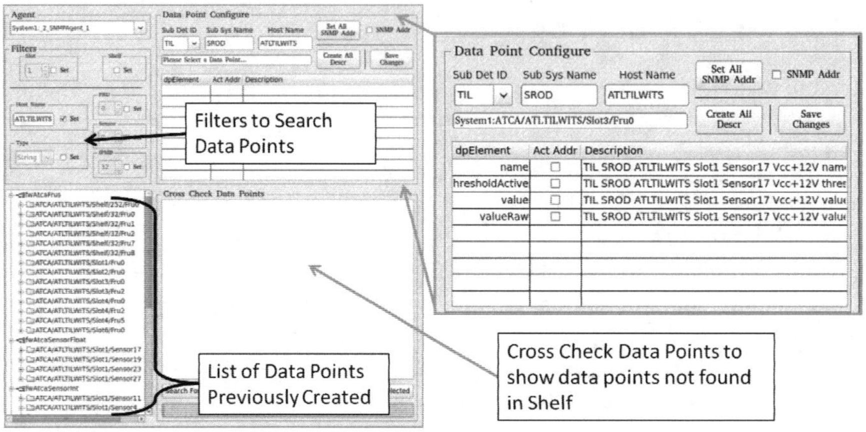

Figure 5: Framework panel that allows data point configuration.

or have been removed from the chassis, searching data points with filters and controlling the SNMP polling.

6. Conclusions

The CERN community will be upgrading to the new ATCA and μTCA standards as a replacement of the current VME technology. The work up to now has facilitated in the understanding of these new systems. The framework described in this proceedings is an expert tool that will assist in the integration of new hardware into the current Detector Control System. The tool provides a platform that allows the development of custom monitoring scripts and panels that can be used by any sub detector in ATLAS. The tool automates searching, sorting and creation of large complicated back-end infrastructure and provides a scalable framework to assist in the integration efforts.

7. Acknowledgements

The University of the Witwatersrand has facilitated in the efforts by purchasing the required hardware to allow hands on development for this work. The National Research Foundation (NRF) for the bursary they provided. The SA-CERN program provided financial assistance for research visits to CERN. I would also like to thank the School of Physics, the Faculty of Science and the Research Office at the University of the Witwatersrand.

References

[1] The ATLAS Collaboration 2008 *Journal of Instrumentation* **3** S08003–S08003 URL `http://stacks.iop.org/1748-0221/3/i=08/a=S08003`

[2] The ATLAS Collaboration 2012 Letter of Intent for the Phase-II Upgrade of the ATLAS Experiment URL `http://cds.cern.ch/record/1502664?ln=en`

[3] Carrió F, Ferrer A, Castillo V, Hernández Y, Higón E, Fiorini L, Mellado B, March L, Moreno P, Reed R, Solans C, Valero A and Valls J A 2013 *ATLAS Note* URL `http://cds.cern.ch/record/1628753?ln=enhttp://cds.cern.ch/record/1628753`

[4] Ballestreroa S, Farthouatb P, Gorinib B *et al.* 2014 ATCA in ATLAS Tech. rep. URL `https://edms.cern.ch/file/1304001/1/ATCA-Backupdoc-rev2.pdf`

[5] PICMG PICMG - About Us URL `http://www.picmg.org/v2internal/aboutus.htm`

[6] Lantzsch K, Arfaoui S, Franz S *et al.* 2012 *Journal of Physics: Conference Series* **396** 012028 URL `http://stacks.iop.org/1742-6596/396/i=1/a=012028`

[7] Siemens 2014 SCADA System SIMATIC WinCC - HMI Software URL `http://www.automation.siemens.com/mcms/human-machine-interface/en/visualization-software/scada/Pages/Default.aspx`

Journal of Physics: Conference Series **645** (2015) 012025 doi:10.1088/1742-6596/645/1/012025

Exploiting Parallelism in the TileCal Trigger System with GPGPU

Marc Sacks
School of Physics, University of the Witwatersrand, Johannesburg 2050, South Africa
E-mail: marc.sacks@cern.ch

Abstract. After the 2022 upgrades, the Tile Calorimeter (TileCal) detector at ATLAS will be generating raw data at a rate of approximately 41 TB/s. The TileCal triggering system contains a degree of parallelism in its processing algorithms and thus presents an opportunity to explore the use of general-purpose computing on graphics processing units (GPGPU). Currently, research into the viability of an sROD ARM-based co-processing unit (PU) is being conducted at Wits University with especial regard to increasing the I/O throughput of the detector. Integration of GPGPU into this PU could enhance its performance by relieving the ARMs of particularly parallel computations. In addition to the PU, use of GPGPU in the front-end trigger is being investigated on the basis of the used algorithms having a similarity to image processing algorithms - where GPU can be used optimally. The use of GPUs in assistance to or in place of FPGAs can be justified by GPUs' relative ease of programming; C/C++ like languages as opposed to assembly-like Hardware Description Languages (HDLs). This project will consider how GPUs can best be utilised as a subsystem of TileCal in terms of power and computing efficiency; and therefore cost.

1. Introduction

A Toiroidal LHC ApparatuS (ATLAS) is a detector in the Large Hadron Collider (LHC) at CERN [1]. It is a general purpose detector used in standard model and beyond the standard model physics [1]. It comprises several sub-detectors; namely the inner detector (ID), the electromagnetic and hadronic calorimeter, and the muon spectrometer [1]. The hadronic calorimeter generates a large amount of data very rapidly: currently 205 Gbps and in 2022 about 41 Tbps [2]. This amount of data requires specialised computational systems in order to be processed. This calorimeter functions in conjunction with a read-out driver (ROD) which is responsible for reconstructing the energy profile of detected events [2]. The ROD is being updated to a super ROD (sROD), this document describes a proposed ARM-processor/GPU based sROD co-processor. Before data reaches the ROD it is processed by systems belonging to the calorimeters, the algorithms involved in these electromagnetic and hadronic calorimeters will be considered in terms of parallel processing [1].

2. The ATLAS Detector

The purpose of the various detectors is to measure the energy and trajectories of the particles formed within it due to high-energy proton-proton (p-p) and heavy ion collisions [1]. The ID reconstructs vertex and momentum measurements for charged particles, the muon spectrometer measures the momentum and position of muons. The final detector, positioned in between the ID and muon spectrometer, is the calorimeter [1]. The calorimeter has two subsystems, namely the electromagnetic (EM) and hadronic calorimeters [1]. The EM detector is a liquid argon based detector, while the hardonic calorimeter is composed of sheets or tiles of plastic scintillators sandwiched between inactive plates of steel. The tiles of plastic scintillator have led to the hadronic calorimeter being referred to as the tile calorimeter or TileCal.

3. Tile Calorimeter

TileCal is approximately 11 m in length and its inner and outer radii are 2.28 and 4.25 m respectively [1]. It is divided into three barrel-like regions; one central barrel and two lateral barrels. The barrels are made up of 64 wedges composed of plates of iron and the plastic scintillator. At each of the tiles' edges are wavelength shifting fibres connected to two photomultiplier tubes (PMTs). It is the signal from the PMTs that form the basis of the TileCal trigger system.

3.1. TileCal Trigger System

The trigger system was developed out of the need to separate rare physical processes from background noise. Proton bunches collide in ATLAS every 25 ns [1]. This 40 MHz bunch crossing rate produces a number of detectable particles which, when converted into a digital signals, generate data at 205 Gbps. After the 2022 upgrade of TileCal, this will increase to 41 Tbps. Storing this amount of data is neither possible nor desirable. Instead the approach is to detect events of interest in real-time and to store those events for analysis - all other data is discarded. The trigger system is divided into a Level-1, Level-2 and Higher Level Trigger (HLT). Level-1 is a rapid hardware-based algorithm which selects events from the 40 MHz bunch crossing in such a way as to reduce the recorded events to approximately 75 kHz. The software based Level-2 trigger and HLT further reduce the recorded event rate to 3.5 kHz and 200 Hz respectively. A system overview is shown in Figure 1. Signals from the calorimeters are digitised and undergo processing which organises data into region of origin and time of origin (and therefore bunch crossing). The data is then sent to the Cluster Processor Modules (CPMs). These modules are responsible for implementing the logic which dictates whether an event will be sent to the read-out driver (ROD) and eventually to the Level-2 trigger and HLT. The logic implemented by the CPMs will be discussed in relation to GPU after a discussion of parallel programming.

4. Parallel Programming Paradigms and GPGPU

The progress of CPU technology has seen a trend of ever faster clock speeds [3]. This of course implies the ability to perform more calculations per second in a serial fashion, meaning calculation *n+1* cannot occur before calculation n has terminated. Due to physical and monetary limitations, clock speeds can no longer be made faster and so to achieve more efficient computing (in terms of power consumption, time and budget) an alternative paradigm must be sought. Parallel programming and processing attempts to increase computational efficiency by processing calculation n+1 while simultaneously processing calculation n (provided the calculations are independent of one another.) Parallel processing is essentially the distribution of a computational load over several processors. These processors can be entire computers, as is the case for a network, or it could be a single integrated chip (IC) possessing multiple cores [3]. The latter is the true for Graphic Processing Units (GPUs), which can possess hundreds of CPUs on a single IC [3]. Each of these cores need not operate as rapidly as a conventional processor to achieve efficient computational results, because the cores operate in tandem to solve multiple calculations simultaneously. GPUs, as the name suggests, were developed with the specific intention of processing and rendering graphics. Initially GPUs were largely inaccessible and used only for their original purpose, but as their usefulness became recognised, they begun to

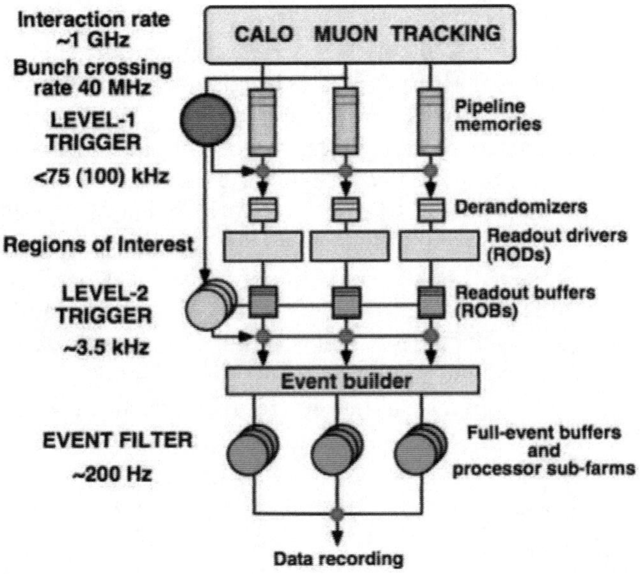

Figure 1. An overview of the ATLAS calorimeter trigger system.

be utilised for processing computational loads other than graphics. To facilitate this, GPU programming languages have been developed which are similar in style to C/C++ and are thus accessible to a large number of software developers. Certain computations are more suited to parallel processing than others. The computations must be large enough to benefit from a distribution over multiple processors, and the calculations must be as independent of each other as possible. If the calculations depend on one another then it could be the case that they cannot run in parallel, hence defeating the purpose of parallel processing.

5. GPGPU Applicability to Calorimeters

The problem then becomes one of assessing how appropriate parallel computing is in the case of the TileCal Level-1 trigger.

5.1. GPGPU and the sROD

The ROD receives data from TileCal and is responsible for reconstructing the energy profiles of events before sending them to the Level-2 and HLT [2]. The energy reconstruction is done using an optimal filtering algorithm based on a weighted sum of the data samples [2]. The main components of the ROD are two FPGAs and two DSPs [2]. For the 2022 upgrade, an updated ROD referred to as the super ROD or sROD is proposed [2]. The current sROD design is FPGA based, however research into an ARM processor based co-processor unit for the sROD is currently being conducted at the University of the Witwatersrand [2]. Apart from performance enhancement, the design is a move from specialised, costly equipment to readily accessible inexpensive equipment. The ARM processors are capable of performing the optimal filtering algorithm, however it is proposed that GPUs can be used in conjunction with ARM processors to form a more robust co-processing unit, especially in light of the fact that the algorithm is not serial in nature.

Journal of Physics: Conference Series **645** (2015) 012025 doi:10.1088/1742-6596/645/1/012025

5.2. Cluster Processor Module Algorithms

The Cluster Processor Module (CPM) algorithm decides whether an event is sent to the ROD [1]. It operates as follows. As stated, the three barrel regions are made up of 64 wedges. Each of these wedges is radially segmented into 3 layers. These 3 layered wedges are referred to as trigger towers. The CPM algorithm considers the trigger towers in sets of 4x4 overlapping windows, one such window it shown in Figure 2. The EM algorithm sums the E_T for each of the four 1x2 and 2x1 combinations as illustrated in Figure 2, and each combination is checked against a programmable threshold. This sum is then compared to the sums for the 12 adjacent windows as indicated in Figure 3 in order to locate a local maximum, this is referred to as the isolation threshold. The same is done with the hadronic algorithm, but each sum is then added to the sum of the 2x2 hadronic calorimeter region. The same isolation threshold is performed. Of importance to this discussion is the sliding-window nature of the CPM module. If one considers each position of the window to be a pixel making up an image, then the algorithm can be thought of as an image processing one. It is thus suggested that GPUs could be used as a replacement to the FPGAs currently in use.

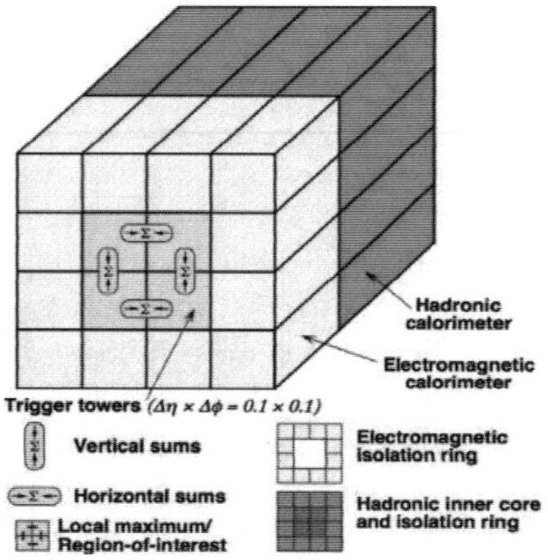

Figure 2. An illustration of the CPM summing process [1].

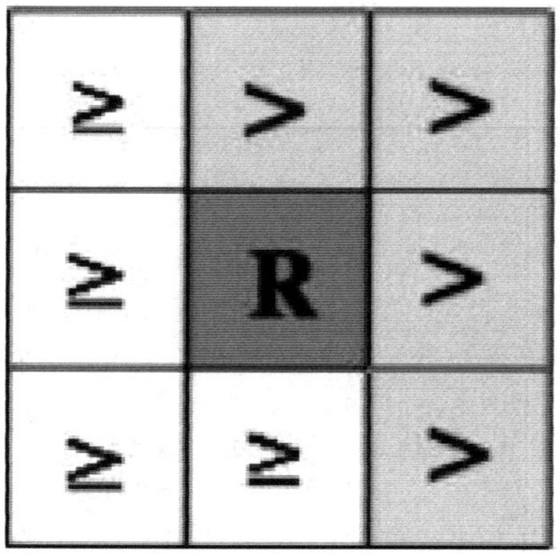

Figure 3. An illustration of the isolation threshold [1].

6. Conclusion

In order to deal with the massive amount of data generated by the hadronic calorimeter after the 2022 upgrade, GPGPU could be used in combination with ARM processors to provide an energy-efficient, user-friendly processing unit.

References

[1] The ATLAS Collaboration *et al* 2008 *JINST* **3** S08003
[2] Cox M, Reed R, Mellado B 2015 *JINST* **10** C01007
[3] Pacheco P S 2011 *An Introduction to Parallel Programming* (USA: Elsevier Inc)

Journal of Physics: Conference Series **645** (2015) 012026 doi:10.1088/1742-6596/645/1/012026

ANA - A framework for building ATHENA on ARM

Joshua Wyatt Smith, Andrew Hamilton

Department of Physics
The University of Cape Town, Rondebosch, South Africa

E-mail: joshua.wyatt.smith@cern.ch

Abstract. The ATLAS software framework (ATHENA) is large and dynamic, comprised of around 6.5 million lines of code. It is compiled using the ATLAS monitoring system, NICOS which uses tools and scripts located and tuned for the CERN services, LXPLUS and AFS. Furthermore, the constraints placed on the hardware that the software is based, limits compilations to traditional x86 architecture. With the sudden interest in ARM processors for large scale high energy physics computing, a new system needs to be implemented to build ATHENA versions for ARM, on ARM. This letter serves to introduce a building framework called Atlas Nightly on ARM (ANA). This new framework implements patches to suit the ARM architecture with the goal of a final ATHENA version for ARM.

1. Introduction

The ATLAS [1] experiment at the Large Hadron Collider (LHC) [2] is one of the experiments responsible for probing Standard Model and Beyond Standard Model physics, using the proton-proton collisions of the LHC. In 2011 and 2012 the center of mass energies reached 7 TeV and 8 TeV, respectively. The LHC has been shutdown for the last year due to upgrades but is expected to restart in March 2015 with a center of mass energy of 13 TeV. The detector's hardware and software have to keep up with this increased flux of data. Furthermore, once these data have been collected, it must be analyzed at server farms around the world. Complex and powerful software has been created for the ATLAS experiment in the form of a "software stack" and is deployed on traditional x86 computers. The type of analysis that ATLAS (and all experiments at the LHC) do is that of high throughput computing. This means that there is no upper bound on the number of CPUs that can be used when analyzing the data. The cost of running these server farms from the perspective of power usage now becomes an important factor.

ARM [3] processors have recently seen a rise in popularity within the scientific community. Their respectable processing speed and very conservative power usage makes ARM an interesting choice in high energy physics computing. With the introduction of the new 64-bit ARM server it is clear that ARM processors are still going through improvements. It is therefore desirable to have a system in place so that the ATLAS software can be repeatedly and reliably compiled for the ARM architecture. The work presented here was carried out on the Nvidia Jetson TK1 development board as a proof-of-concept. The board has a quad-core cortex-A15 processor with a maximum power draw of approximately 12 Watts, running Fedora 21.

2. The ATLAS software

The software stack used by ATLAS, Athena (named after the Greek goddess of wisdom) is complex and dynamic. It thus is more practical to separate the different packages into individual projects and then build from ground up. Fundamental projects known as LHC Computing Grid (LCG) software, Gaudi [4] and ATLAS specific code comprise an Athena release. A release consists of around 6.5 million lines of code. Roughly 4.5 million lines are written in C++, 1.5 million in python, while other languages make up the rest. Athena versions are compiled using the NIghtly COntrol System (NICOS) which facilitates the building of the individual packages and projects for such a large framework. Complete versions are compiled every night for the x86_64-slc6-gcc48 platform (i686-slc6-gcc47 and x86_64-slc6-gcc47 are also compiled for older versions of Athena) with standard tests being carried out simultaneously. Fixes are performed the next day once bug reports have been generated. Releases are compiled using *CMT* [5] (a package managing tool), however, there is a migration to using the faster more flexible tool, *CMake*. This will enable the ability to cross-compile for the ARM architecture.

Figure 1 shows the software structure for an Athena build (read from bottom to top), specifically for release 20.1.0. This is the release chosen to be compiled for ARM where the projects in orange were neglected in the build. These focus more on "online" computing whereas our focus will be "offline" computing. The LCG software is comprised of common packages that are experiment independent. This includes around 200 packages such as Python, Boost and ROOT. The LCG software and Gaudi were compiled natively on the Jetson-Tk1 board using the *CMake* project. The rest of the projects (from DetCommon and up) were compiled using CMT in a new framework called ATLAS Nightly on ARM (ANA). This tool will be explained in the next section.

3. ANA

The objective of ANA is to be able to repeatedly and consistently build versions of Athena for the ARM architecture with minimal input from the user. In these early stages of development it's not feasible to go through the entire ATLAS code looking for issues that may or may not appear. The approach taken with ANA is to build the packages while taking note of what fails. The developer then writes a patch that can then be applied or reversed. This way, records are kept of what changes are made and in the case of incorrect diagnoses this can be easily corrected.

It is important to note that only the ATLAS code compiled using CMT has been

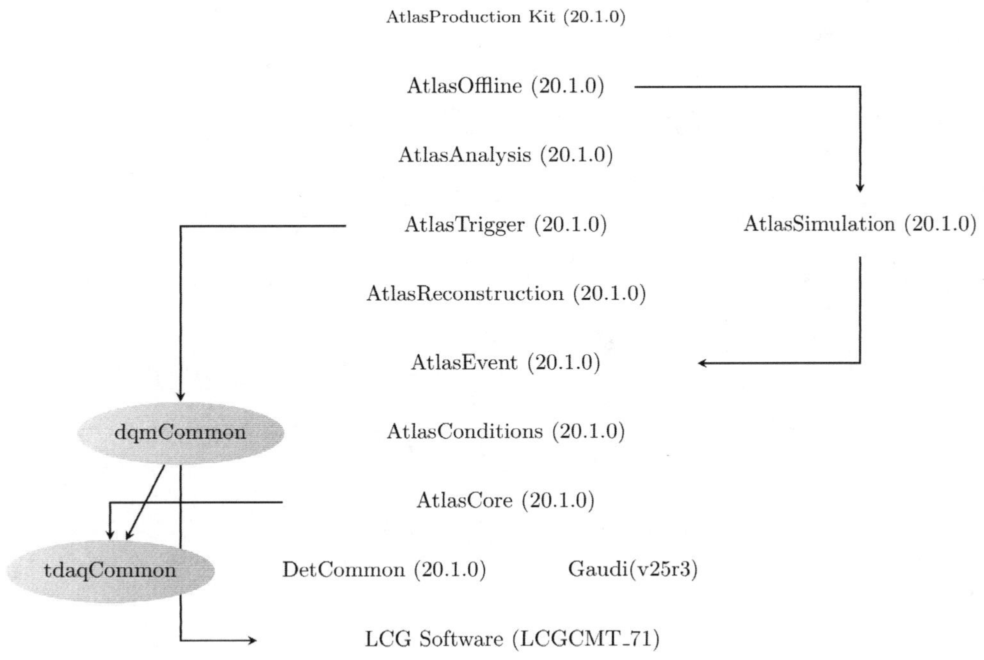

Figure 1. The ATLAS software hierarchy with an experiment independent base, LCG software and Gaudi.

included into ANA version 1.0. The means that the LCG software and Gaudi need to be compiled beforehand and still require essential manual tweaks to accommodate the ARM architecture. For example, one of the more crucial tweaks is to define the new ARM architecture in the LCG_Platforms *requirements* file. In this case it means creating the tag *armv7l-fc21-gcc49-opt/dbg* (the *CMTCONFIG* variable, which translates to the armv7l architecture on Fedora 21 using GCC 4.9) and defining it as Linux.

The structure of ANA is shown in Figure 2 with an explanation of the more important files/directories shown on the figure.

To run ANA one needs the source code located at `https://github.com/jwsmithers/AtlasOfflineBuild-framework`. Once this has been downloaded all that is required is to set the environment variables in *Environments.sh* and then do

```
>> source BuildAtlasOffline.sh.
```

When this command is run it creates the correct CMT directory structure, *project.cmt* and *requirements* files, and gives the option to pull the required SVN packages based on the *VERSION* variable set earlier on (provided this hasn't been done before, it is a lengthly process). ANA then provides the options to either apply all the patches in the *patches* directory or skip them, as well as choosing the project that you want to build (or by default building them all). Around 65% of the ATLAS code has been ported to ARM using ANA with the rest closely following. Some basic example scripts (such as "AthExHelloWorld") have run successfully.

Journal of Physics: Conference Series **645** (2015) 012026 doi:10.1088/1742-6596/645/1/012026

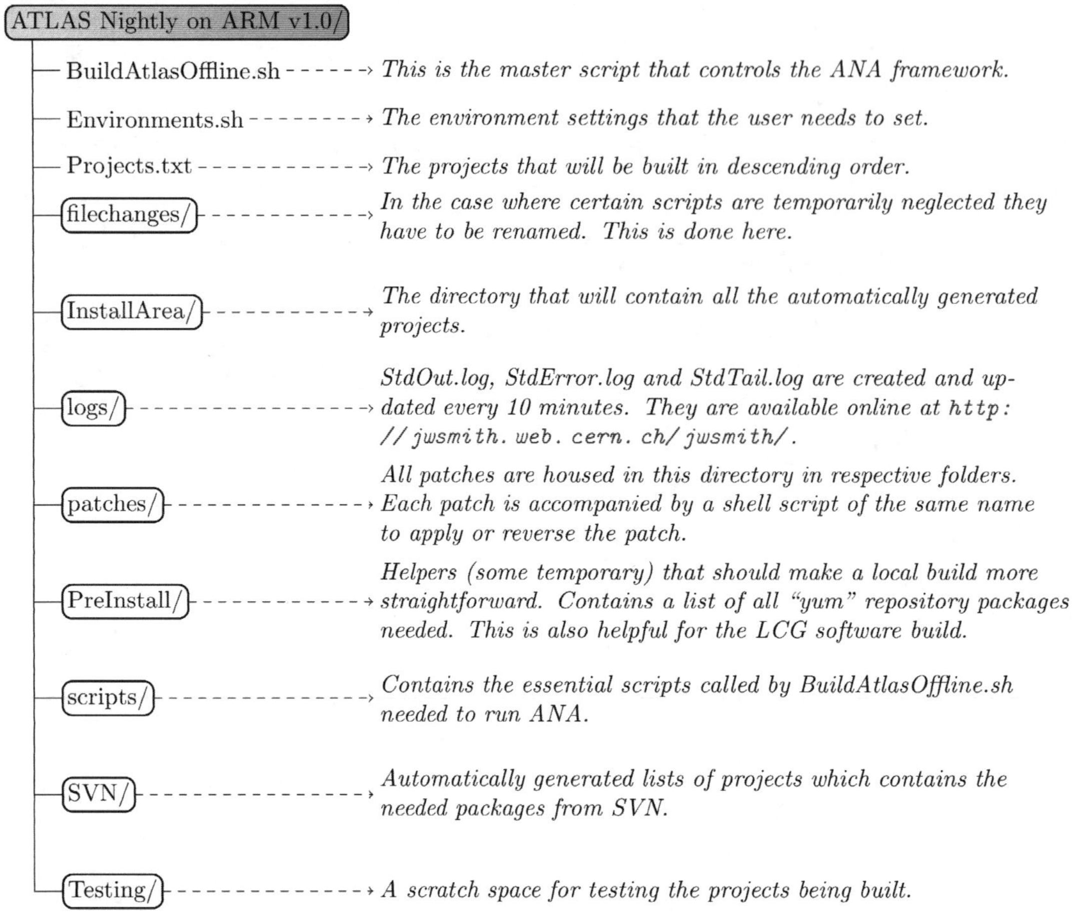

Figure 2. An explanation of the directory structure of ANA.

4. ARM related issues

Issues arose when compiling on ARM architecture. This first few apply to LCG software:

- ROOT-v34-21: A patch written by a team from CMS [6] was used that enabled the building of Cintex. A further patch was written to enable ROOT to compile on Fedora 21.
- Grid specific packages: None of these packages can be compiled as they are specific to the x86 architecture. For now, these are not essential packages.
- Oracle DB: There are no ARM 32-bit builds and Oracle doesnt plan to provide them. It is still being determined how much of a setback this will be.

For the ATLAS software most of the patches add the new ARM related tags into the code. They also fix path related errors due to the software being optimized for tools found in AFS and compiled for x86 architecture. There are still some oddities that need to be mentioned:

- Syntax errors: A few syntax errors were found such as missing "=" or "::" signs.

Journal of Physics: Conference Series **645** (2015) 012026 doi:10.1088/1742-6596/645/1/012026

- Linking errors: Some libraries are linking in the wrong order and thus fail the first time being compiled. The temporary solution is to do a "make" twice.
- *Genreflex* compiler flag issues: The compile flags *-fPIC*, *-fsigned-char* and *-fsigned-bitfields* don't get removed when the *lcgdict* macro is called. Thus, it sees it as passing an option and so fails. The solution is to add private macros which remove these options every time *lcgdict* is called. This could stem from a tag being defined incorrectly.
- Virtual memory use: Occasionally a package would fail to compile because it exhausts the virtual memory address space. A temporary fix is to just recompile the package.

These highlight some major fixes. There are many more which are all documented in the *patches* directory of ANA.

Summary

With the sudden and gaining interest in the ARM processor we felt it became prudent to build a framework in which stable and reliable ARM builds can take place overnight. The ANA framework builds the ATLAS software stack with minimal input from the users side. Developers also have a clearer view into the layout of ANA and how it functions should they want to make contributions. Currently, around 65% of the entire software stack has been ported to the ARM architecture.

Aknowledgements

The authors would like to thank the National Research Foundation of South Africa and the University of Cape Town for funding this research.

References

[1] ATLAS Collaboration, G. Aad *et. al.*, *The ATLAS Experiment at the CERN Large Hadron Collider*, *JINST* 3 (2008) S08003.
[2] L. Evans and P. Bryant, *LHC Machine*, *JINST* 3 (2008) S08001.
[3] "Arm Cortex-A series." `http://www.arm.com`.
[4] G. Barrand *et. al.*, *GAUDI - A software architecture and framework for building HEP data processing applications*, *Comput.Phys.Commun.* 140 (2001) 45–55.
[5] C. Arnault, *Configuration Management Tool*, 2006.
[6] CMS Collaboration (Chatrchyan, S. et al.), *The CMS experiment at the CERN LHC*, 2008.

Journal of Physics: Conference Series **645** (2015) 012027 doi:10.1088/1742-6596/645/1/012027

Development of a high data-throughput ADC board for the PROMETEO portable test-bench for the upgraded front-end electronics of the ATLAS TileCal

Matthew Spoor, Oscar Kureba and Charles Sandrock

School of Physics, University of the Witwatersrand, Johannesburg, South Africa

E-mail: matthew.spoor@cern.ch, oscar.kureba@cern.ch, Bruce.Mellado.Garcia@cern.ch

Abstract. The Large Hadron Collider (LHC) is preparing for a major Phase-II upgrade scheduled for 2022 [1]. The upgrade will require a complete redesign of both on- and off-detector electronics systems in the ATLAS Tile hadron Calorimeter (TileCal) [2]. The PROMETEO (A Portable ReadOut ModulE for Tilecal ElectrOnics) stand-alone test-bench system is currently in development and will be used for the certification and quality checks of the new front-end electronics. The Prometeo is designed to read in digitized samples from 12 channels simultaneously at the bunch crossing frequency while accessing quality of information in real-time. The main board used for the design is a Xilinx VC707 evaluation board with a dual QSFP+ FMC (FPGA Mezzanine Card) module for read-out and control of the front-end electronics. All other functions are provided by a HV board, LED board and a 16 channel ADC daughter board. The paper relates to the development and testing of the ADC board that will be used in the new Prometeo system.

1. The ATLAS Tile Calorimeter

The ATLAS (A Toroidal LHC ApparatuS) experiment is a general purpose particle detector at the LHC at CERN. The experiment is designed to record proton-proton collisions occurring at a bunch crossing frequency of 40 MHz. At such high frequencies large amounts of raw data are generated leading to data flows in the Tbps range. The ATLAS experiment comprises of several sub-detector systems each aimed at recording different aspects of a collision.

The TileCal is the central hadronic calorimeter sub-detector in ATLAS (Figure 1a.) [3]. TileCal comprises of four barrel sections, two central barrels (LBA and LBC) and two extended barrels (EBA and EBC). Each section is divided into 64 azimuthal slices with each slice being further segmented by a lattice of alternating steel plates and plastic scintillating tiles (Figure 1b). When particles travel trough TileCal they collide with steel plates causing them to break up into showers of charged and neutral particles. As these particles travel through the scintillating tiles small amounts of energy is deposited which generates photons. The photons are picked up by wavelength shifting optical fibres which carry the signal to the front-end electronics found within the outermost part of the sub-detector. Photomultiplier tubes (PMTs) are used covert these faint light pulses into analog electrical signals. These signals are then amplified, digitised and stored in temporary pipeline buffers while they wait for trigger selection. Only samples of these events that are selected by the L1 trigger are sent to the back-end electronics for further

Journal of Physics: Conference Series **645** (2015) 012027 doi:10.1088/1742-6596/645/1/012027

processing. After three levels of trigger selection the data flow is reduced to a few 100 Mbps which is stored for off-line analysis.

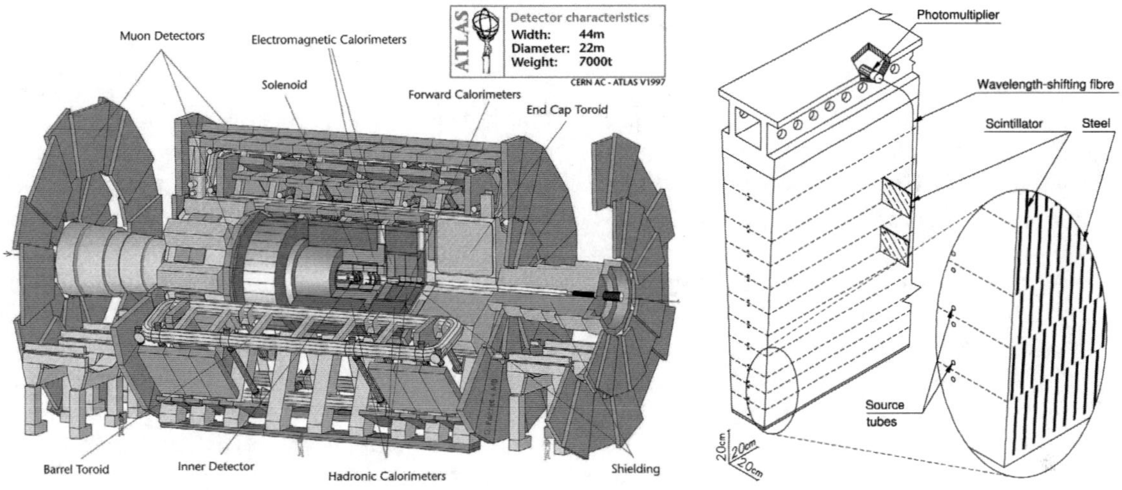

(a) (b)

Figure 1: a) ATLAS detector and b) TileCal module.

2. Phase II upgrade

The Phase II upgrade will increase the design luminosity of the LHC by a factor of 5 to 7. [1] The ATLAS experiment will need to undergo some substantial changes if it is to meet these demands. A complete redesign the detector electronics will be needed in order to meet increased radiation tolerances and higher data processing requirements [4]. The current read-out electronics uses pipeline memories in the font-end to store digitised samples while they wait for trigger selection. After the upgrade the read-out system will transmit all data directly to the back-end using gigabyte optical links during every bunch crossing. This will provide digitally calibrated information with enhanced precision and granularity to the first level trigger in order to improve both trigger latency and resolution.

The current read-out hardware is located in 2 m long "superdrawers" within each module of the TileCal. [5] The read-out electronics systems in each superdrawer are daisy-chained together forcing them to share the same data connection to the back-end and same power supply. A new front-end architecture has been proposed that breaks each superdrawer down into 4 "minidrawers". This architecture will merged several of the old boards onto a more compact 3 board system that will include massive levels of redundancy to increase the reliability of the system. Each minidrawer will be completely independent of the others, having its own powers supply and optical links to the back-end.

3. Prometeo

The new front-end electronics being developed for the Phase II upgrade will need a compatible test-bench system. The PROMETEO (Portable ReadOut ModulE for Tilecal ElectrOnics) is being developed to replace the existing MobiDICK [6] test-bench which has been in operation service since 2003. Prometeo (Figure 2a) is a stand-alone test-bench system that will be used for the full certification of the new front-end electronics of the ATLAS TileCal. The TileCal has 256 modules each one will need to be independently serviced and evaluated using the Prometeo portable test-bench. All tests must be completed during LHC shut-downs to make sure the sub-detector is ready for data-taking periods. The Prometeo is high data-throughput system required to process all the information produced by one minidrawer in real-time. A test setup needs to be able to analyse data from 12-PMTs at LHC bunch crossing frequency assessing the quality of the data in real time and diagnose malfunctions [7].

3.1. Hardware Design

When designing the system it was important that it was easy to upgrade when components become obsolete but still remain relatively cheap to build [7]. The motherboard used in Prometeo is the Xilinx VC707 evaluation board which contains a Virtex 7 FPGA. The board was chosen for it high IO capabilities and ability to process large amounts data in real-time. Several other boards and modules are able to connect to the board through multiple connections. A high speed QSFP+ module attaches to one of the two FMC connectors. It is used for the communication between the minidrawers and test-bench. The other FMC connector supports a custom ADC board which was developed to digitise the trigger signals coming from the front-end adderboards. Additionally there is a HV board to provide power to the PMTs, a LED driver board to inject light pulses into the PMTs during tests, a router used to communicate to the users and a power supply unit to power the test-bench. The VC707 is programmed with the IPBUS protocol which is capable of storing trigger data samples which can be retrieved upon a trigger request

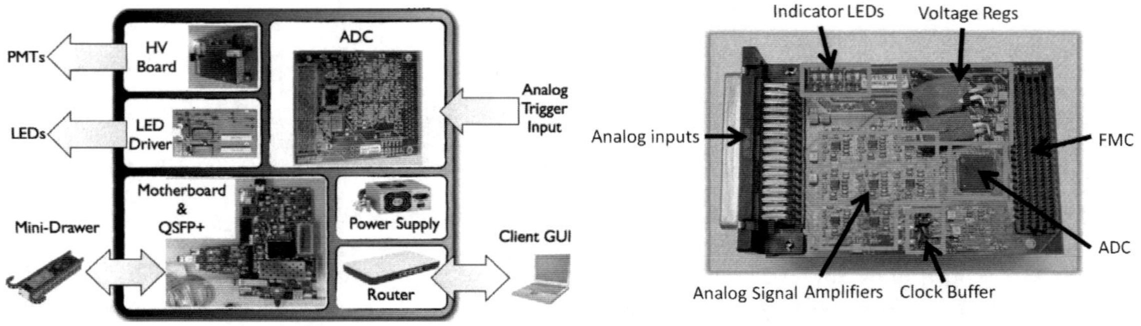

(a) (b)

Figure 2: a) Prometeo Overview b) Prometeo ADC board.

4. ADC daughter Board

The ADC daughter boards (Figure 2b) function is to digitise analog trigger signals coming from the front-end electronics adder-cards [7]. The board communicates with a FPGA located on the VC707 through a high speed FMC connection. The board has two high-performance ADC571 chips each with 8 differential input channels used data sampling. The ADCs have a 12-bit resolution and are capable of sampling data at a rate of 50 Mega Samples a Second (MSPS).

In the Proemteo system the FPGA sends a 40 MHz clock to a clock buffer chip on the ADC board which cleans the signal and sends it to both the ADCs. This clock is used by the ADCs as a reference for data sampling. Several Different voltage regulators are used on the board to improve signals integrity and remove signal coupling problems. Indicator LEDs on the board inform the user that all the voltages are at the correct level. Analog data being sent to ADC board is first conditioned and filtered before being passed to the ADCs for digitisation. ADCs can then digitise and serialise the data from each channel and send it to the FPGA. A 240 MHz bit clock is also created by ADCs and sent alongside the data. This clock is used is used for the de-serialisation of the samples. Once the serial data arrives in the FPGA it is de-serialised and stored in temporary buffer in the VC707 RAM. If data is seen to be interesting it will be called from the buffers and further processed otherwise it will be overwritten. Each differential channel coming from the ADCs transports data at 480 Mbps. There is 16 channels in total coming from the ADC board leading to a total data transfer rate of 7680 Mbps.

4.1. Firmware
Custom firmware was developed for the FPGA so it can control and manage the data read-out of the ADC board. Firmware was described using VHSIC Hardware Description Language (VHDL) using Xilinx ISE design software. The firmware for the Prometeo is highly modular which enables different sections to be tested and used independently. As the same ADC chips were used in both MobiDick and Prometeo, similar algorithms could be used for the data read-out.

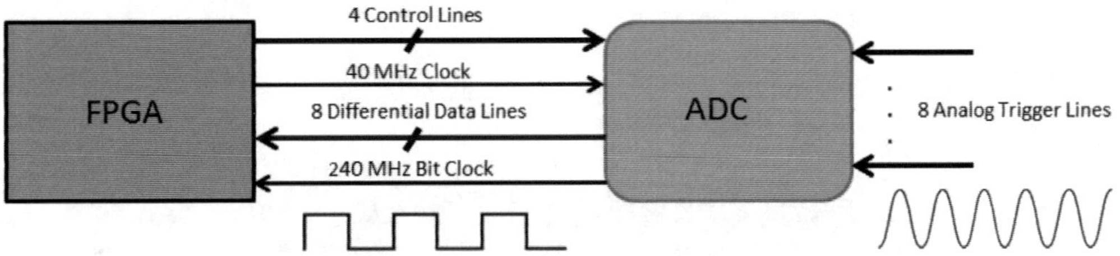

Figure 3: Data signals between an ADC and FPGA

Figure 3 shows all the data connections needed between a single ADC and the FPGA. Each one of these signals needs to implemented and simulated within firmware. Initial testing of the ADC board involved configuring the ADCs to output the test pattern 10101010 (Deskew) from each channel. To do this serialised commands are sent from the FPGA through the control lines using a set of strict timing requirements in order to configure the ADC to send the pattern. These pattern tell us if ADCs are receiving our commands, if they are sending data at the correct speeds and if the 8 data lines are synchronised correctly. Figure 4a shows the basic order of events when doing this test.

4.2. PCB
The first prototype for the new ADC board was manufactured in 2014 in South Africa. The board went through several tests to see if it performed as expected. Chipscope IP cores were used in the firmware to allow viewing of the data signals between the FPGA and ADC board during operation. This kind of testing allows the developer to confirm that the firmware and the hardware is interacting as designed. The Figure 3b shows actual commands being sent to the ADC board to configure it to send the Deskew patten. The top 8 line show the 4 commands lines of both ADCs while the bottom 6 show the high speed data being sent back.

After testing the two prototype ADCs boards at the WITS High Throughput Electronics Laboratory many design faults could be identified. The circuit design and PCB layout of the board could be modified and improved for the next generation of ADC board. Significant changes included adjusting the power regulators, reducing noise through component and wire shifting

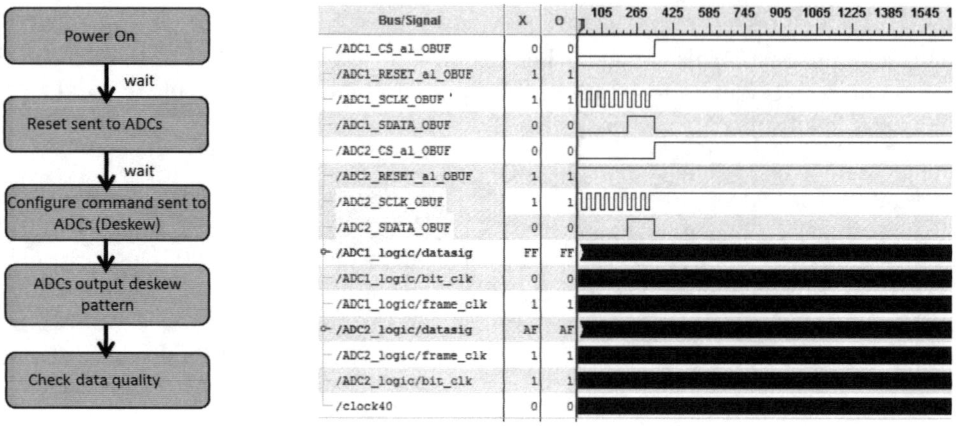

(a) (b)

Figure 4: a) Basic testing flow b) ADC board setup using Chipscope.

and the addition of a 10 Pin connector to control the LED driver board. The Figure 5 below shows the new PCB layout of the next ADC board iteration.

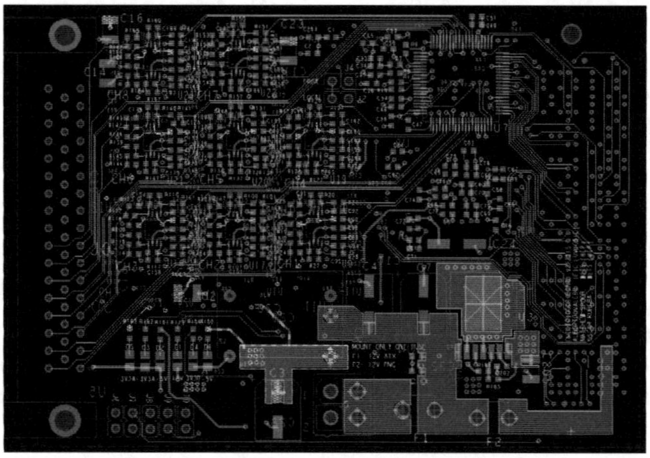

Figure 5: New ADC board PCB Layout

5. Conclusion

The Prometeo is in the process of being designed for the certification of the front-end electronics for the Phase-II upgrade. The high throughput system will be capable of performing all the testing needed during shut-down periods. The ADC board is still under going further testing before it will ready for use on the Prometeo system. The next generation of the board has be designed is currently being manufactured in South Africa. Once the new board is produced final firmware testing can be completed and the board will be certified for use.

Acknowledgments

I would like to thank the University of the Witwatersrand, University of Valencia and the SA-CERN program. The TileCal upgrade collaboration for significant contributions to this work. I would also like to thank the School of Physics, the Faculty of Science and the Research Office at the University of the Witwatersrand. Finally I would like to thank my supervisor Bruce Mellado Garcia.

References

[1] Collaboration A 2012 Letter of Intent for the Phase-II Upgrade of the ATLAS Experiment *CERN-LHCC-2012-022*

[2] Carrio F 2014 Upgrade of the ATLAS Tile Calorimeter Electronics *ATL-TILECAL-PROC-2014-002* URL `http://cds.cern.ch/record/1702969`

[3] Collaboration A 2008 *Institute of Physics Publishing and SISSA* 6 URL `http://iopscience.iop.org/1748-0221/3/08/S08003`

[4] Carrió F, Castillo V, Ferrer a, Fiorini L, Hernández Y, Higón E, Mellado B, March L, Moreno P, Reed R, Solans C, Valero a and Valls J a 2014 *Journal of Instrumentation* **9** C02019–C02019 ISSN 1748-0221 URL `http://iopscience.iop.org/1748-0221/9/02/C02019`

[5] Ajaltouni Z, Badaud F and Bouhemaid N 1994 *ATLAS Internal Note* 1–27 URL `http://inspirehep.net/record/1195072/files/tilecal-94-043.pdf`

[6] Alves J, Kim H Y, Minashvili I, Moreno P, Reed R, Schettino V, Solans C, Souza J, Usai G and Valero A 2013 A new portable test bench for the ATLAS Tile Calorimeter front-end electronics certification *ATL-TILECAL-PROC-2013-005* pp 1–6

[7] Bullock D, Carrió F, Govender M and Hofsajer I 2014 *Proceeding of Science* **ATL-TILECA** 6–9

The data acquisition system for a fixed target experiment at NICA complex at JINR and its connection to the ATLAS TileCal readout electronics

K G Tomiwa[1*], I Slepnev[2] and S Bazylev[2]

[1] School of Physics, University of the Witwatersrand, Johannesburg 2050, South Africa
[2] Joint Institute of Nuclear Research, Dubna, Russia

E-mail: `kehinde.gbenga.tomiwa@cern.ch`

Abstract. Today's large-scale science projects have always encountered challenges in processing large data flow from the experiments, the ATLAS detector records proton-proton collisions provided by the Large Hadron Collider (LHC) at CERN every 50 ns which results in a total data flow of 10 Pb/s. These data must be reduced to the science data product for further analysis, thus a very fast decisions need to be executed, to modify this large amounts of data at high rates. The capabilities required to support this scale of data movement is development and improvement of high-throughput electronics. The upgraded LHC will provide collisions at rates that will be at least 10 times higher than those of today due to it's luminosity by 2022. This will require a complete redesign of the read-out electronics and Processing Units (PU) in the Tile-calorimeter (TileCal) of the ATLAS experiment. A general purpose, high-throughput PU has been developed for the TileCal at CERN, by using several ARM-processors in cluster configuration. The PU is capable of handling large data throughput and apply advanced operations at high rates. This system has been proposed for the fixed target experiment at NICA complex to handle the first level processes and event building. The aim of this work is to have a look at the architecture of the data acquisition system (DAQ) of the fixed target experiment at the NICA complex at JINR, by compiling the data-flow requirements of all the subcomponents. Furthermore, the VME DAQ modules characteristics to control, triggering and data acquisition will be described in order to define the DAQ with maximum readout efficiency, no dead time and data selection and compression.

1. Introduction

High energy physics experiments are confronted with the challenge of handling the volume of data recorded by particle detectors. One of such detector is the ATLAS Detector of the Large Hadron Collider(LHC), which records proton collision at about 50ns resulting to data output rate of about 10 Pb/s, with the anticipated upgrade of the detector to about ten times its previous recording energy which is about 6.5TeV (13 TeV collision energy), the data output rate would increase [1].

The TileCal detector which is a sub-detector of the ATLAS detector used to measure energy and position of hadrons [2], the scintillating tiles of the detector produce light when particles from collision crosses them, light are converted to analogue signal by the photo-multipliers and to digital signals by digitizers. The particles bunch crossing frequency in ATLAS experiment results in millions of particles event per seconds and in turns a large volume of digital signals.

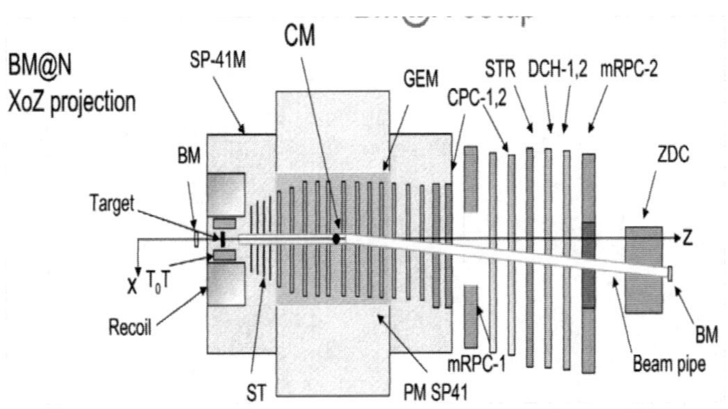

Figure 1. BM@N Setup (XoZ projection)[3].

In order to handle this data, high throughout electronics are developed. These electronics are high performance system, capable of selective reading and performing basic reduction and data filtering. A Processing Unit with 4 advanced micro X-Gene system on chip (SoC) was has been developed to either operate as a standalone device or a supporting device to other FPGA based system (sROD) in ATLAS experiment.

The Baryonic matter at Nuclotron (BM@N) project in the Nuclotron-Based Ion collider Facility (NICA) at the Joint Institute of Nuclear Research (JINR); is an experiment proposed to the study of AU+AU collision, with beam energy of up to $\sqrt{S_{NN}} = 11GeV/n$ (Giga electron volts/particle) [3]. The proposed particle detectors (the BM@N detector Fig. 1) will be used to record particle produced by the fixed target experiment, by combining high precision track measurement with time-of-flight information for particle identification and total energy measurement for event characterization.

The BM@N detector has different sub-detectors, strategically positioned around targets to make different observations of the fixed target experiment. According to the BM@N letter of intent the BM@N setup will contain approximately 370000 channels of the front-end electronics, the sub-detector front-end electronics pre-amplify signals generated from observations. These signals are sent to the data acquisition (DAQ) system for digitization, compression and selection by high-throughput electronics (PU). The PU developed for TileCal in the ATLAS experiment will be used for First level processes (data check, flow control, zero suppression and integrity check), sorting distribution and Event buffering.

2. General purpose ARM-based Processing Unit

The challenge posed by High Volume throughput computing (HVC) and desire to handle this challenge with cost effective technology has led to the development of a processing unit Fig. 2 (schematics) from low cost ARM system on chips (SoCs). The PU design with several ARM processors in cluster configuration, this arrangement enhance the performance of the PU to be able to process at least 40 GB/s (Gigabits/s) of raw data fed through the I/O interface (PCI-Express , XAUI or bonded SFP+Connectors) [4]. This PU can be used for different high-level functions on the high-throughput raw data processing, such as spectral analysis and and online histogram monitoring.

3. BM@N Detector overview

Fig. 1 shows the schematic setup of the experiment, the setup combines different sub-detectors to observer different physics phenomena of the fixed target experiment.

Journal of Physics: Conference Series **645** (2015) 012028 doi:10.1088/1742-6596/645/1/012028

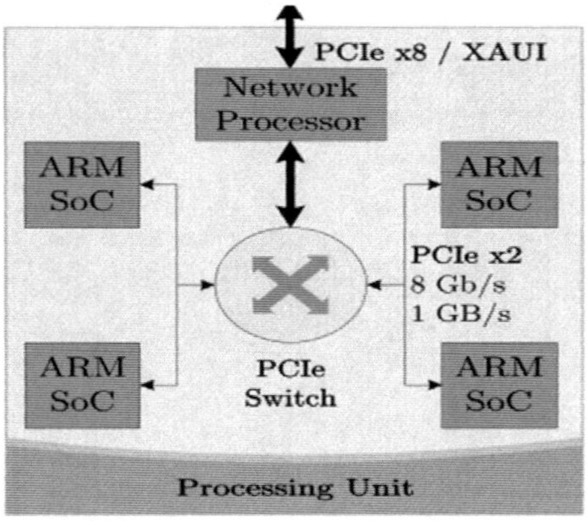

Figure 2. Schematics of the PU [4].

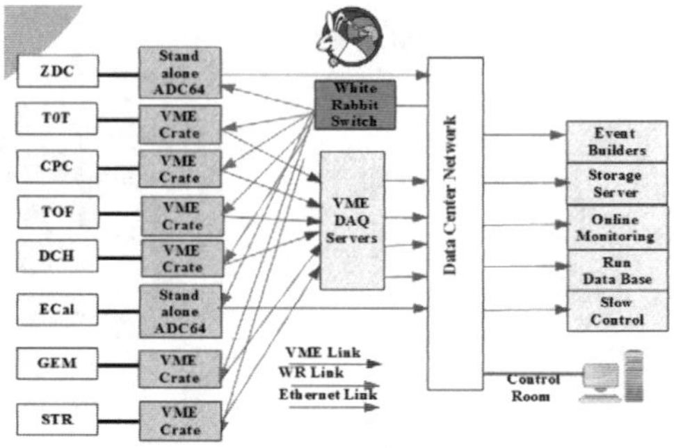

Figure 3. BM@N DAQ System.

- The Gaseous Electron Multipliers (GEM) detectors has twelve planes located downstream of the target inside the analyzing magnet (0.8T magnetic field) are designed to measure charge track multiplicity for particle identification.
- The time-of flight detector (TOF) are based on multigap resistive plate chamber (mRPC). The set up uses two TOF: the outer and inner TOF (mRPC-1) detector and the near TOF detector (mRPC-2).
- Fast Start detector (T_0T) is installed close to the target, the detector is designed to form T0 and L1 centrality trigger. The (T_0T) is a Cherenkov detector with time resolution of about $30ps$, it can be combined with the multiplicity counter (MC) to L0-triggr signal.
- The Straw tubes (ST) for intermediate tracking.
- Drift chamber (DC) the setup uses two DC planes situated outside the magnetic field, they are use as intermediate trackers.
- Zero Degree Calorimeter (ZDC) is designed for collision centrality analysis by measuring the energy of forward going particle. it helps with central event selection at trigger level during data taking.
- Cathode pad chamber (CPC) is one of the tracking systems, it provides additional precise space coordinates of particles based on center of gravity measurement.
- Electromagnetic Calorimeter (ECal), designed to measure the spatial position and energy of particles in the fixed target experiment.

4. Data Acquisition and Trigger system overview

The BM@N trigger and data acquisition system (TDAQ) (Fig. 3) can be decomposed into three logical parts: the DAQ subsystem itself, the fast control and the trigger (FC+TU) subsystem: which intent to produce and handle the synchronous signals and the slow control (SC) system. For simplification of the interaction between DAQ elements, provision of standalone working mode for each sub detectors, easy reconfiguration of active sub-detector and easy hardware replacement, the DAQ hardware units are functionally subdivided into hierarchy by logical level along the data stream. The levels of the DAQ system comprise the Front End module of each sub-detector (FEM),Sub-event builder of subcomponents (SubEvent), Event Buider (EvB), Pool level and storage level.

4.1. Brief Description of the VME modules

- **ADC:** ADC are the waveform digitizer, they take continuous instant samples of input pulse signal at a fixed rate and output a complete digital representation of the signal. It is a 64-channel 12-bit with signal processing core module, with 62.5 Million sample per event with and Ethernet interface, it is designed to perform data readout by single fibre-optical link and capable of time and obtain synchronization by the white Rabbit switch.
- **TDC:** The time to digital converter is a time interval instrument; it is also know as a Time-Of Flight digitizer. it records the time-stamp of input pulses of particles. The main features are time stamping, multihit and trigger matching capability. The module is based on HPTDC chip developed at CERN, with a VME64X interface and VMEDAQ TTC bus (clock, trigger).
- **TQDC:** The TQDC is an FPGA based module, a multi-hit time-stamping TDC and also a waveform digitizer. it is used to measure pulse arrival time, charge and purse shape.
- **FVME2:** FVME2 is a 6U Master for the Data Acquisition system, the module automatically readout other VME module in chained mode. It is majorly a VME system controller with SFP serial link (800 Mbps or 2.5 Gbps M-Link).
- **FVME2TM/FVME2TMWR:** FVME2TM is the Trigger, Timing and Control Module for the VMEDAQ system, it has onboard time generator. It takes spill and trigger input used for data acquisition.
- **CTF:** CTF6 is the central trigger distributor equipped with white rabbit node core, the module is designed with 1 to 6 clock (125MHz LVDS clock) and trigger.

4.2. Data Acquisition system

The BM@N experiment uses the VME Data acquisition system standard, with the capability of run control, online data display, instant status monitoring and testing, automatic detection and configuration of modules. The VME crates host the BM@N readout module (ADC, TDC and TQDC), the VME controller modules (FVMEs) and the clock and trigger modules (FVMETMs). The BM@N modules digitize analogue signals (waveform and time) from dedicated sub-detectors, they store data in buffer for read out to the VME server over the VME link, The VME server is connected to the data network via 1Gb Ethernet link.

The central data center network hosts a general purpose processing unit which is capable of implementing data check, flow control, zero suppression and event building.

4.3. Trigger system

The FVME2TM/WR are the VME crates trigger modules, they have onboard 41.667 MHz clock generator used for trigger and time control. They receive input for spill and trigger from the central trigger processor and distribution, spill signals indicate the start of event, it is used as a gate for the data acquisition.

The LI trigger (Fig. 4) is a low latency signal (about 5 to 10 microseconds), it is distributed by CTF modules, generated by L1 trigger processor(CTF L1TP) and transmitted via high speed serial links. L1 trigger is required for modules equipped with ASICs (HPTDC), it is used to initiate a process that time-stamped and move data entries in modules to their buffers for L2 trigger. The L2 trigger (Fig. 5)signals are high latency digital signals (about 10 to 100 microseconds), implemented over the white rabbit network, used for selective readout from devices buffer.

5. Data Structure

Fig. 6 shows the Raw data format of the BM@N DAQ system. Raw data is stored as its is generated from the hardware; the minimal quantum of data is 4 bytes (32 bits). In the data

Journal of Physics: Conference Series **645** (2015) 012028 doi:10.1088/1742-6596/645/1/012028

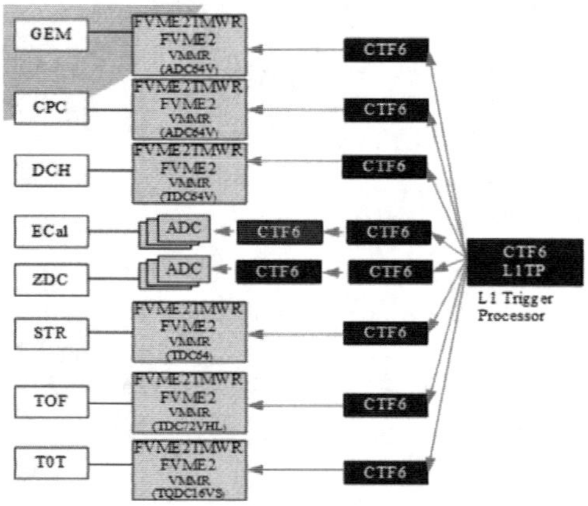

Figure 4. L1 trigger

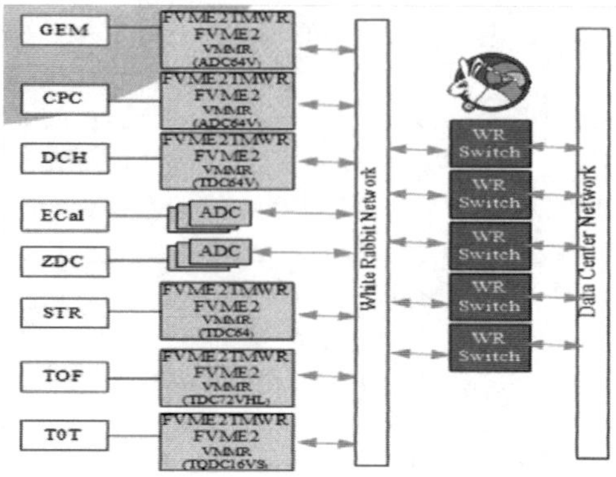

Figure 5. L2 trigger

stream raw data are split into spills, beginning with spill header (SHDR) and ending with spill trailer (STRL). The data in between the SHDR and STRL are sequence of zero or more events. Event begins with event header (EHDR) and ends with event trailer (ETRL), the data between EHDR and ETRL are sequence of module data block. Module data also starts and ends with module header (MHDR) and module trailer (MTRL), module data is any number of words of data, each module has their specific raw data format with module header comprising of slot number and module ID used to identify the slot and specifies the module a set of data is coming from.

The stand alone ADCs use M-stream waveform digitizer data format, data contain channels being read out in current event. The M-stream produces two types of data: the subtype 0 and subtype 1. The subtype 0 consist of a mask of channels being read out, they are used for zero suppression and basic reduction during data processing while the subtype 1 consist of waveform digitized data fragments as its payload.

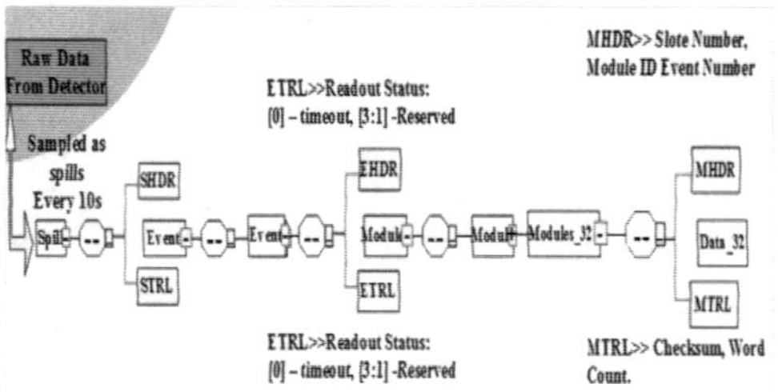

Figure 6. Raw data format

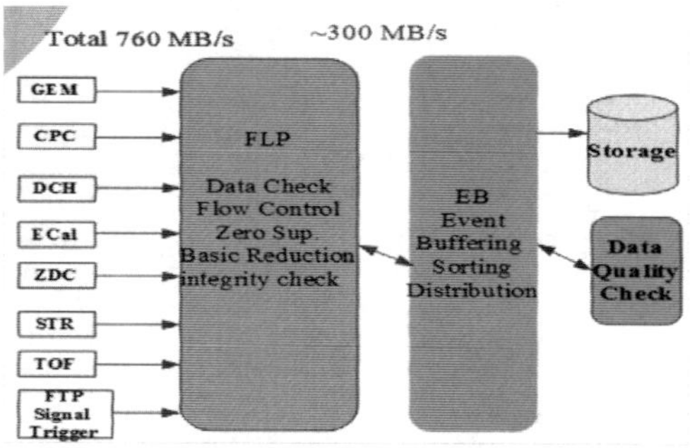

Figure 7. Data flow

6. BM@N data flow and event size

Fig.7 shows the DAQ system data flow, analogue signals from each sub-detector are amplified and digitized by dedicated front-end electronics. Raw data from waveform digitizer (standalone ADCs) are read uncompressed at high speed as compared with time digitizers (TDCs and TQDCs) to the first level process (FLP). The first level process is to reduce data flow by eliminating unimportant data, integrity check and zero suppression with nontrivial algorithms.

Journal of Physics: Conference Series **645** (2015) 012028 doi:10.1088/1742-6596/645/1/012028

Table 1. BM@N Event size for 2017.

Detector	Readout card	Geometry	DAQ CH	occupancy	Bytes/ch	Bytes/Event
CPC	ADC64V	2*9129/64	285	3.84	4	4400
DHC	TQDC64V	2*2048	4096	0.06	4	1500
ECAL	ADC64	9*360	3240	0.30	44	44000
GEM	ADC64	96000/64	1500	3.34	4	23000
STR	TDC64V	2*6*450	5400	0.06	4	1700
TOF	TDC72V	2*2*1536	6144	0.06	8	3700
ZDC	ADC64	104	104	0.03	44	1400
Trigger	TQDC16V		1	1	100	100
Others			1	1	500	500
Total						80,000

A high performance PU is being developed for this purpose, the PU would be equip with 4 advance micro X-Gene system on chip (SoC) processors to enhance the performance.

Table 1 shows the estimated event size of 2017 BM@N experiment, data harness at trigger rate of 10kHz.

7. Conclusion

The total data rate from the frontend modules to FLP is about 760 MB/s, after the first level processing data would be reduced to 300MB/s at event building. Because of the rate of volume of data an high-throughput PU will be developed to handle data reception, flow control, integrity check and zero suppression at high speed with no dead time.

Acknowledgments

The authors would like to thank Vladimir Sidorenko and all the members of AFI group of the Joint Institute of Nuclear Research Dubna, Russia. The School of Physics University of Witwatersrand Johannesburg, South Africa.

References

[1] Peter V 2012 ATLAS Upgrade for the HL-LHC: meeting the challenges of five-fold increase in collision rate *arXiv:1201.5469v1 [physics.ins-det]* B **1** 1

[2] Carrio F, Castillo V, Ferrer A, Fiorini L, Hernandez Y, Higon E, Mellado B, March L, Moreno P, Reed R, Solans C, Valero A and Valla J 2013 *ATLAS Collaboration JINST 9 (2014) C02019* **2013-020**

[3] *http : //nica.jinr.ru/files/BM@N/BMN_CDR.pdf.*

[4] Cox M, Reed R, and Mellado B, 2015 The development of a general purpose ARM-based processing unit for the ATLAS TileCal sROD *JINST C01007* **10**